Mathematics for the Middle Grades (5-9)

1982 Yearbook

Linda Silvey
1982 Yearbook Editor
Sepulveda Junior High School
Los Angeles, California

James R. Smart
General Yearbook Editor
San Jose State University

National Council of
Teachers of Mathematics

Libary of Congress Cataloging in Publication Data:

Main entry under title:

Mathematics for the middle grades (5–9)

(Yearbook ; 1982)
Bibliography: p.
1. Mathematics—Study and teaching (Secondary)
—Addresses, essays, lectures. I. Silvey, Linda.
II. Smart, James R. III. National Council of
Teachers of Mathematics. IV. Series: Yearbook
(National Council of Teachers of Mathematics) ; 1982.
QA1.N3 1982 [QA12] 510s [372.7] 81-22524
ISBN 0-87353-192-2 AACR2

Printed in the United States of America

Table of Contents

Part III: Games, Contests, and Student Presentations

Preface

Mathematics for grades 5–9 is particularly appropriate as a topic for an NCTM yearbook because (1) grades 5–9 are crucial years in the mathematical development of students; (2) grades 5–9 are also critical years in the developmental process of students as they mature from children to young adults; (3) the emergence of the organization of the middle school has focused attention on the special needs of students and their teachers in the middle grades; (4) each of the NCTM recommendations for school mathematics of the 1980s, contained in *An Agenda for Action* and supported by *Priorities in School Mathematics,* has special significance in grades 5–9; (5) although these grades have been recognized as vital, relatively little professional material with particular emphasis on mathematics curriculum has concentrated on these grades alone; and (6) whereas some new and worthwhile ideas have appeared in the mathematics curriculum for these grades, these are not usually presented as a unique collection of instructional activities.

Recognizing the importance and appropriateness of focusing on this topic, the Educational Materials Committee of the NCTM selected mathematics for grades 5–9 as the theme of the 1982 Yearbook. As indicated by the list of authors who wrote the articles in this book, a dedicated body of individuals is represented, including classroom teachers, supervisors, and teacher educators, concerned with the improvement of mathematics instruction in grades 5–9. It is hoped that the content in this yearbook will serve to supplement and reinforce activities and strategies appropriate for the middle grades.

The articles in this yearbook are presented in three major sections. The first section focuses on critical issues in mathematics in grades 5–9. The first article introduces the yearbook and some of the major contributions of middle grade mathematics to the school curriculum. From the many possible issues that could have been selected, those chosen for discussion include sex-related differences in mathematics instruction, teaching mathematics to learning disabled students, computer literacy, and numerical and arithmetical concepts with emphasis on problem solving.

The second section of the yearbook features articles that discuss various learning activities including mathematics indoors, outdoors, in the community, and during the Roaring Twenties. Also included are activities and worksheets for teaching geometric concepts, using the calculator, working with decimals, and beginning equation-solving procedures. Unusual approaches, suitable for large-group, small-group, or laboratory type of instruction, are described by suggestions for using newspapers, sports cards, statistics represented graphically, learning centers, and various problem-solving techniques.

The last section discusses games, contests, and student presentations, starting with three articles offering ideas for game strategies and game development. Another article suggests how to use games to organize a tournament or contest, and the final two articles offer suggestions for organizing student carnivals and fairs where middle grade students can demonstrate their knowledge of, and interest in, mathematics.

Ninety-seven proposals for articles were submitted for possible publication in this yearbook. The task of reviewing the articles, selecting those to be included, and making suggestions for the revision of those selected was carried out by the editors and a dedicated and knowledgeable advisory committee. Committee members read each manuscript at each stage of submission. Our thanks and appreciation go to the advisory committee members:

Sheila Berman, Patrick Henry Junior High School, Los Angeles, California

Stanley J. Bezuszka, S.J., Boston College, Chestnut Hill, Massachusetts

Robert Hamada, Instructional Planning Division, Los Angeles Unified School District, Los Angeles, California

Margariete Montague Wheeler, Northern Illinois University, De Kalb, Illinois

The editors further wish to thank the NCTM Educational Materials Committee and the NCTM headquarters staff for their interest in this topic and their guidance through the process of planning and producing the book. We also wish to thank each author of an article appearing in the book and all those who submitted proposals or manuscripts for consideration.

It has been our pleasure to prepare this yearbook. We believe it will serve as a sourcebook for new activities, techniques, ideas, and background material. We sincerely hope that these materials will foster a continued improvement in the teaching of mathematics in grades 5–9.

LINDA SILVEY
1982 Yearbook Editor

JAMES R. SMART
General Yearbook Editor

I. Critical Issues

1

Middle Grade Mathematics: An Overview

Margariete Montague Wheeler
Stanley J. Bezuszka, S.J.

CONSIDER the following ordered triples:

3-5-4	4-5-3	5-2-5	6-4-2
4-4-4	5-3-4	6-2-4	7-3-2
4-3-5	5-4-3	6-3-3	8-1-3

For the purposes of the 1982 Yearbook of the National Council of Teachers of Mathematics, each of these ordered triples represents one of the school organizational patterns for grades 1 through 12 that have actually been in operation during the past nine decades—the late 1800s and all of this century. The school between the elementary school on the one side and the senior high school on the other has at several points in time responded with varying degrees of emphasis to demands arising from laws and regulations from the community and understandings, observations, and opinions of professional groups. The various school organizational patterns represent more than an effort to shuffle grades; they are a deliberate response to pressures arising from laws regulating child labor, compulsory school attendance, teacher specialization, and teacher competence. They are also a response to broader understandings of aptitude and attitude differences among early adolescents, observations of earlier physical and emotional maturation among students, and opinions of professional and community groups concerning the content of a general education curriculum.

Changing societal values in conjunction with changing curricular emphases brought about shifts in emphasis concerning the function of what has been variously called the intermediate school, the junior high school, and the middle school. Unfortunately, the name on a particular school building is not always indicative of the prevailing educational philosophy within. It is not always clear whether the school organizational pattern was designed to facilitate the students' transition to the subject-centered curriculum of the senior high school (the junior high school organizational pattern) or to acknowledge the special cognitive, sociological, and physiological characteristics of the early adolescent (the middle school pattern).

The focus of this yearbook, however, is not on the genesis of the educational unit that receives students from the elementary school and sends them into the senior high school. Rather, the focus is on the mathematics program that spans all or part of this educational unit: the mathematics program of the middle grades—grades 5 through 9.

Seldom has the early adolescent, the student found in the middle grades, been the subject of intense scrutiny. As one educational psychologist noted from a 1965 perspective (Wattenberg 1965, p. 39):

> The extent of our ignorance is symbolized by the fact that during the past four decades, years during which 50,000 books a year on the most detailed and even inconsequential subjects of all kinds have appeared in the United States, there have been exactly three books devoted to pre-adolescence, the pivotal psychological state of the junior high school population.

More recently, within the last five years, the absence of cohesive bodies of knowledge concerning the early adolescent as a learner of the sciences was detailed in a paper commissioned by the National Science Foundation (Haertel 1978). Haertel repeatedly noted the need for coordination between our knowledge of the formal reasoning abilities of the early adolescent and the sequencing and teaching of middle grade mathematics and science. She developed a strong argument for the need for further study of cognitive abilities and of the relationships between cognitive style and instructional practices.

Accumulated research seems to indicate that the cognitive abilities available to an individual are different at different points in the life cycle and at different points in adolescence. These include the ability to represent (model), to classify, to form concepts, to memorize, to solve problems, and to reason. At the risk of overgeneralization, it can be noted that virtually all students at early adolescence seem to have attained what Piaget termed the concrete operational stage, whereas formal abilities, when exhibited, are specific rather than general and are, perhaps, influenced by instruction.

Though the task of developing and teaching a mathematics program for the middle grades has undoubtedly presented many teachers and members of the professional educational community with opportunities for thinking about the early adolescent as a learner, the purpose of this yearbook is not to describe the tumultuous, transitory period known as early adolescence. (Readers wishing to pursue these points are referred to Johnson [1980].) Rather, contributors to the yearbook acknowledge that there are many different things that could be done in middle grade mathematics programs and, indeed, that many different things are being done. After identifying several of the critical issues facing teachers of middle grade mathematics, this yearbook—in distinct, though interdependent, chapters—deals with what is being and what could be accomplished.

To be an effective teacher of middle grade mathematics is to be an individual responsive to a variety of mathematical requirements and pressures from both school and nonschool sectors of society. These influences include but are not limited to—

- the preservation of mathematics as an important component of our scientific culture;
- the development of future consumers of mathematics, be these consumers sociologists, tool and die makers, physicists, linguists, marine engineers, insurance adjusters, or dieticians;
- the recognition and encouragement of mathematical talent, despite awesome variations in individual differences among the students;
- the development of users of elementary mathematical techniques, including the ability to express relationships in a variety of ways, to compute numerically, to solve a broad range of problems, to reason abstractly, and to evaluate results. (The most recent position paper itemizing these skills and techniques is *An Agenda for Action: Recommendations for School Mathematics of the 1980s* [NCTM 1980].)

Care must be taken that pressures such as these are not directed solely at the teacher of middle grade mathematics. That is, the burden of these requirements cannot be assumed to exist only for the middle grade teacher of mathematics and not for colleagues at the elementary, high school, or university levels. Concomitantly, it cannot be assumed that the distribution of such pressures is uniform across the grade levels. It will be argued here, however, that the impact of these mathematical pressures is different for the teacher of middle grade mathematics.

In many instances the body of arithmetic, informal geometry, and introductory algebra commonly labeled middle grade mathematics culminates a student's study of mathematics. Beyond the end of the middle grades, a student must elect either to continue the study of mathematics or not to study mathematics at all. The implicit and explicit pressures arising from the terminal aspects of a bounded curriculum frequently give rise to an emphasis on the mastery of a narrowly defined collection of *basic skills.* As a consequence, it is incumbent on the teacher of middle grade mathematics to avoid what a former president of the National Council of Teachers of Mathematics has called a "barren curriculum . . . a review and minor extension of the mathematics of grades one through six" (Smith 1978, p. 84). Without arguing barrenness as a cause or as an effect, many educators would identify the intent of mathematics programs in the middle grades as an extension of the understandings and skills of mathematics already learned and an exploration of new areas of mathematics. The burden is different for teachers of middle grade mathematics because all too often what is intended is remedial teaching of the arithmetic of the elementary school and readiness teaching for the study of algebra.

Seldom are teachers in the middle grades certified to teach all the grades from 5 through 9. This contrasts sharply with their colleagues from the elementary school and the senior high school who are certified to teach across the entire range of grades within that school. In most instances, a middle grade teacher has specialized either in elementary school curriculum or in a secondary school content area, but not in both. As a consequence, the middle grade mathematics teacher with a specialization in elementary education may have had from one to four semesters of mathematics or mathematics methods or an undergraduate major or minor in mathematics. These classroom teachers are specialists in the rather clear-cut mathematics programs of the elementary school or the senior high school but not in the mathematics program of the middle grades. They are not prepared to handle both a different treatment of familiar content and a treatment of content intrinsically satisfying to mathematicians.

Mathematics for the Middle Grades (5–9) is partitioned into three distinct, related sections. The first section deals with current, critical issues of mathematics instruction in the middle grades. Although the issues identified in the first seven articles should not be construed to be exhaustive in scope or equivalent in importance, they do suggest that diversity among learners and among competing curricular emphases contribute to an unevenness in middle grade mathematics programs. In school mathematics, the mathematics being taught, the methods by which that content is taught, and the learner to whom it is taught are highly complex entities that interact in ways not yet fully understood. Although it is beyond the scope of this yearbook to relate these issues to four recent calls for improvement in school mathematics (MAA 1978; NACOME 1975; NCSM 1978; NCTM 1980), it is appropriate that selected issues be addressed further:

- *Do middle grade girls have the same opportunities to learn mathematics that boys have?*

Fennema closely examines several variables that help in understanding why the seemingly equal opportunities of middle grade girls and boys to learn mathematics do not result in equal outcomes. She describes specific actions to help the classroom teacher facilitate the learning of mathematics by all students.

- *What intervention techniques facilitate teaching mathematics to the middle grade child with special needs?*

Ten alternatives for teaching mathematics to the learning disabled student are detailed by Bley and Thornton. Several of the alternatives are highlighted in an instructional sequence for writing fractions.

- *In what ways does mental arithmetic or computer literacy enhance the middle grade curriculum?*

Two articles focus on different reasons for expanding the middle grade curriculum. Thompson notes in one that middle grade students are, and will continue to be, living, working, and playing in a computer-oriented society. These students must therefore possess a general understanding of the capabilities and limitations of computers. Thompson offers a plan for integrating computer literacy into existing mathematics programs. In the other article, Atweh argues that although mental arithmetic can never displace paper and pencil as a computation aid, there are important reasons for expanding the curriculum to include the skills of mental arithmetic.

- *Must students experience difficulties understanding rational number concepts?*

Post, Behr, and Lesh integrate recent findings from psychological, mathematical, and instructional research to present insights into the problems involved in teaching rational number concepts and processes in particular and the learning of mathematics in general. The detailed analysis of rational number found in this article is an important review of a body of mathematics that is included to a varying degree in each of the middle grades.

- *Can more than lip service be given to the development of problem-solving skills in middle grade students?*

Three teachers from West Virginia answer this question affirmatively. To support their position, Charles, Mason, and White present classroom-tested ideas for teaching behaviors before, during, and after a student has worked with a problem.

An assumption that mathematics teaching in the middle grades can be an invigorating adventure permeates the two remaining sections of the yearbook. With varying degrees of specificity, thirty authors in twenty-one chapters encourage mathematics teachers of the middle grades to develop a repertoire of alternative teaching strategies. These include organizing and implementing mathematics fairs, developing and using instructional games, strengthening problem-solving skills, and using students' daily experiences to capitalize on the great variety of mathematics outside the textbook and outside the four walls of the classroom and to develop motivational settings for applications of mathematics previously learned.

The Council's recommendations for school mathematics of the 1980s, *An*

Agenda for Action (NCTM 1980), contains a strongly worded position that the concept of basic skills in mathematics must encompass more than computational facility. To clarify which skills are basic, the *Agenda* cites the ten basic skills previously identified by the National Council of Supervisors of Mathematics: problem solving; applying mathematics in everyday situations; alertness to the reasonableness of results; estimation and approximation; appropriate computational skills; geometry; measurement; reading, interpreting, and constructing tables, charts, and graphs; using mathematics to predict; and computer literacy. Each of the fifteen chapters in the activities section of this yearbook addresses one or more of these skills.

Pervading the literature of mathematics education is the observance of the difficulties of integrating and coordinating mathematics instruction with instruction in other school subjects. The call for a wide variety of applications of mathematics previously learned is widespread, although the quantity and quality of materials for middle grade classrooms is minimal. (Additional resource material can be found in two recent publications of the National Council of Teachers of Mathematics [MAA/NCTM 1980; NCTM 1979].) Teachers of middle grade mathematics looking to integrate the mathematics they teach with other fields of endeavor should find additional ideas in the articles of this yearbook dealing with the decade of the 1920s (Burnett), mathematics beyond the four walls of the classroom (Casey [with Sowell]), mathematics from sports trading cards (Kuhl), opinion polls (Vance), and statistics through graphics (Sanok).

Middle grade students have only a limited number of strategies from which to draw when solving routine algebra problems. Whitman argues the need for an intuitive approach to equation solving as well as the need for more traditional approaches. Her argument, and that of Atweh in an earlier article, is not for the creation of a false dichotomy (intuition versus formalism) but for diverse instructional strategies.

Where does a classroom teacher find the data for real-world problems? One resource is the daily newspaper. Sowell (with Casey) provides a variety of exercises and activities focusing on such topics as averages, ratios, large numbers, and percentages. Finding routine applications of mathematics in everyday settings is clearly possible.

In part because geometry-rich programs seldom exist, the *Overview and Analysis of School Mathematics, Grades K–12* report (NACOME 1975, p. 146) calls for new and imaginative approaches to the study of geometry in the middle grades as well as in the elementary and secondary grades. In its position paper on the ten basic mathematical skills, the National Council of Supervisors of Mathematics included as one of them the learning of geometric concepts needed to function in a three-dimensional world (NCSM 1978). For readers taking the position that more than a minimal knowledge of a minimal amount of geometry is desirable, three articles should prove to be of particular interest: "Spatial Visualization" (Lappan and Winter), "Math

Mapping Begins at Home" (Hatchett), and "Graph Paper Geometry" (Burger).

Recommendations that mathematics programs take full advantage of the power of calculators and computers at all grade levels challenge the traditional curriculum and instructional priorities of the middle grades. Fisher and Jones have addressed this challenge in an article dealing with "messy" arithmetic problems ("Large Numbers and the Calculator"). A companion article by Lichtenberg and Lichtenberg ("Decimals Deserve Distinction") notes that many punches on a simple four-function calculator give rise to decimals. The authors use the calculator and the monetary system to develop and reinforce decimal concepts.

Problem solving is at the top of the Council's *Agenda.* In the middle grades, solving routine problems builds on the skills developed in the elementary school; solving nonroutine problems encourages flexibility in the approaches used. Three articles on problem solving emphasize a consideration of the procedures used to arrive at the solution of a problem. Questioning whether the information provided is sufficient and whether all the information given is necessary is one procedure considered. Krulik discusses the importance of analyzing the information provided and suggests learning centers as a means to this end. Scalzitti draws an analogy between the strategies needed for solving a crime and those needed to solve routine story problems. In this highly motivational setting, she encourages middle grade students and their teachers to ask: What information is given? What is unknown? What is needed? In the third of these related articles, Reeves considers problems that are not resolved by processes leading directly to a solution. With careful attention to classroom implementation procedures, he encourages using nonroutine problems and opportunities for creative approaches toward a solution.

Should games be used to teach fraction concepts and skills, scientific notation, or the evaluation of algebraic expressions? Evidence from a variety of sources indicates that students and their teachers perceive games to be useful for learning mathematics. Because perception is not sufficient as a rationale for using a game, authors of two articles detail evidence from research that games can be an effective means for teaching selected middle grade skills and concepts. Rules, sample game boards, and suggestions for modifications are provided for fraction games (Bright and Harvey), scientific notation games (Collins), and algebra skill games (Ilani, Taizi, and Bruckheimer).

"Let's have a math tournament! A math carnival! A math fair!" The practicalities of accepting such a challenge can be awesome. Six mathematics educators with a variety of classroom experience in New York, Michigan, and California point the way for such an undertaking. In three articles, they outline the decisions to be made, the projects that students might undertake, a timetable, and the benefits that students and teachers can derive. Del

Regato raises and answers questions that must be considered when organizing a math games tournament. Eschner and Krist design a mathematics carnival to strengthen mathematics topics previously studied. Day, McNichols, and Robb describe the mathematics activity cards needed for a fair that encourages students to explore, hypothesize, and generalize. Indeed, middle grade students *can* successfully complete projects that require them to go beyond textbook exercises. Their teachers can use diverse instructional strategies to achieve such goals.

The 1982 Yearbook of the National Council of Teachers of Mathematics, *Mathematics for the Middle Grades (5–9),* is a survey for all teachers of mathematics—a survey of issues and ideas, of activities and applications, of decisions and dilemmas. The teachers, the students, and the mathematics curriculum for this span of grades have frequently been neglected except as appendages for the elementary or the senior high school grades. This yearbook is one attempt to fill that gap.

REFERENCES

Haertel, Geneva D. "Literature Review of Early Adolescence and Implications for Science Education Programming." In *Early Adolescence: Perspectives and Recommendations.* NSF Report SE–78–75. Washington, D.C.: National Science Foundation, 1978.

Howard, Alvin W., and George C. Stoumbis. *The Junior High and Middle School: Issues and Practices.* Scranton, Pa.: International Textbook Co., 1970.

Johnson, Mauritz, ed. *Toward Adolescence: The Middle School Years.* Seventy-ninth Yearbook of the National Society for the Study of Education. Chicago: University of Chicago Press, 1980.

Mathematical Association of America (MAA). *PRIME—80: Proceedings on Prospects in Mathematics Education in the 1980s.* Washington, D.C.: The Association, 1978.

Mathematical Association of America and the National Council of Teachers of Mathematics, Joint Committee. *A Sourcebook of Applications of School Mathematics.* Reston, Va.: The Council, 1980.

National Advisory Committee on Mathematical Education (NACOME). *Overview and Analysis of School Mathematics, Grades K–12.* Washington, D.C.: Conference Board of the Mathematical Sciences, 1975. Available from the National Council of Teachers of Mathematics.

National Council of Supervisors of Mathematics (NCSM). "Position Statement on Basic Skills." *Mathematics Teacher* 71 (February 1978): 147–52.

National Council of Teachers of Mathematics (NCTM). *An Agenda for Action: Recommendations for School Mathematics of the 1980s.* Reston, Va.: The Council, 1980.

———. *Applications in School Mathematics.* 1979 Yearbook. Edited by Sidney Sharron. Reston, Va.: The Council, 1979.

Ross, A. M., A. G. Razzell, and E. H. Badcock. *The Curriculum in the Middle Years.* Schools Council Working Paper 55. London: Evans Bros., Methuen, 1975.

Smith, Eugene P. "Strengthening Mathematics Programs for Early Adolescents." In *Early Adolescence: Perspectives and Recommendations.* NSF Report SE–78–75. Washington, D.C.: National Science Foundation, 1978.

Wattenberg, William W. "The Junior High School—a Psychologist's View." *NASSP Bulletin* 49 (1965): 39–44.

2

Girls and Mathematics: The Crucial Middle Grades

Elizabeth Fennema

THE National Council of Teachers of Mathematics has committed itself to the principle that "girls and women should be full participants in all aspects of mathematics" (NCTM 1980). Many believe that such a statement is totally unnecessary, since females appear already to have the same opportunities as males to learn mathematics. Girls and boys are in the same classes, taught by the same teachers in elementary and middle schools. Girls appear to have the same opportunities to elect high school mathematics courses that boys have. Postsecondary educational opportunities are apparently as accessible to females as they are to males. Yet for some reason, or set of reasons, these seemingly equal opportunities do not result in equal outcomes for males and females. Females are currently learning mathematics less adequately than males and are not participating in mathematics-related careers nearly as much as males are.

At least three things are evidence that females achieve less in mathematics: differential enrollment by girls and boys in advanced mathematics courses in high school, lower achievement by females in high-level mathematical tasks, and more negative attitudes on the part of females toward the learning of mathematics. Although middle school girls and boys do not have the option to elect or not elect mathematics, the learning and teaching of mathematics in the middle school has a direct effect on achievement and attitudes toward mathematics. The purpose of this chapter is to point out where sex-related differences in learning mathematics exist and then to explore some variables that may help us understand why, in many cases, females are receiving inadequate mathematics education.

Until about 1972 it was accepted almost without question that, overall, males were better than females at mathematics, starting at least as early as kindergarten. For example, Glennon and Callahan (1968, p. 30) stated that "the evidence would suggest to the teacher that boys will achieve higher than girls on tests dealing with mathematical reasoning while the girls will achieve higher than boys on tests of computational ability." An equally prestigious review of research (Suydam and Riedesel 1969, p. 129) stated that there are no significant differences between the sexes in arithmetic achievement before seventh grade but that boys surpass girls after seventh grade. Suydam and Weaver (1970, p. 4) stated: "In general, boys scored higher in mathematical reasoning and girls were better in fundamentals, though some conflicting evidence has been presented." Aiken (1971, p. 203) wrote that "sex differences in mathematical abilities are, of course, present at the kindergarten level and undoubtedly earlier." One could conclude from these reviews that although some authors believed that sex differences in mathematical achievement did not always appear, most felt that when there was a difference, it was in favor of boys.

After about 1972, several things happened that have made us examine what was believed at that time. Published reports were examined more critically. It became clear that many studies that reported higher male achievement had failed to control the number of courses girls and boys enrolled in during high school. The Fennema-Sherman (1978) studies reported that when girls and boys studied the same mathematics courses during high school, few sex-related differences in achievement were found. The influence of society as a whole on female failure (or success) in mathematics became increasingly clear. Sells (1973) reported on how few women entered college with the necessary mathematics prerequisites for most college majors. The Ford Foundation disseminated information compiled by Ernest (1976) about negative female attitudes toward mathematics.

In addition to these events within the mathematics education community, examinations of the role of women in American society have been made. A portion of those examinations have been concerned with aspects of the educational preparation of females that either enable them to enter, or prohibit them from entering, a variety of occupations and professions. Within the last few years, a tremendous amount of money, time, and energy has been spent on various investigations concerned with women and mathematics.

Although in 1982 sex differences still remain in a variety of mathematics-related variables, the magnitude of each of those differences is usually not large and appears to be diminishing. Several beliefs about sex-related differences are currently being discussed widely. Among them are these:

- Females are not as good at math as males.
- Females lack confidence in learning math.
- Females attribute their own success in math to others.
- Females don't believe math is useful.
- Females don't take advanced mathematics courses in either high school or college.

And, in fact, each of these statements contains some truth. However, each is true only when large groups of females and males are being compared. They are not true for many individuals.

In 1979, in fourth-year high school mathematics classes, there was about a 2:3 girl-to-boy ratio. However, under 10 percent of all boys took fourth-year math. Almost as many boys as girls did not take fourth-year mathematics. In most achievement data, the mean differences between the sexes were usually small. Obviously, many girls were achieving at much higher levels than many boys. Intrasex differences are, in reality, much larger than intersex differences. What each person concerned with the education of girls and boys must do is examine individual achievement, not group achievement. So with the warning of being careful about generalizing from group (female) behavior to individual (specific girl) behavior, let us examine sex-related differences in achievement and attitudes toward mathematics.

Sex-related Differences in Learning Mathematics

The second mathematics assessment of the National Assessment of Educational Progress (NAEP II) provides the most comprehensive look at sex-related differences in learning mathematics on a nationwide basis (Fennema and Carpenter 1981). Information about courses taken and achievement in specific content areas (number and numeration, variables and relations, geometry, measurement, and other topics) and at different degrees of complexity (knowledge, skill, understanding, and application) was gathered from a representative sample of over 70 000 9-, 13-, and 17-year-olds. No clear pattern of differences in achievement between girls and boys was apparent at ages 9 and 13. Girls tended to score slightly higher than boys on the less complex tasks, whereas boys tended to score slightly higher on the understanding and application exercises. At age 17, the average performance of boys exceeded that of girls at every complexity level.

NAEP II compared the achievement of 17-year-old girls and boys who reported that they had been enrolled in the same mathematics courses. For each course background grouping (algebra 1, geometry, etc.), boys' achievement exceeded that of girls. The magnitude of the difference increased consistently in relation to the amount of mathematics taken. In other words, the difference in performance between 17-year-old girls and boys was smallest for students who *had not* taken first-year algebra, and the difference was greater for students who *had* taken first-year algebra. It was even greater for students who had taken geometry, and the trend continued through courses beyond second-year algebra. The achievement differences between girls and boys increased as the cognitive complexity of the items increased. There were smaller differences at the less complex levels and larger differences at the higher levels.

Other studies have confirmed these results (Armstrong 1980; California

Assessment Program 1978). When a large sample of girls and boys are selected from many schools and given tests of mathematics performance, little or no differences in achievement are found in elementary school, but by late high school differences are often found, with boys achieving at higher levels, particularly on content of some sophistication, like problem solving. Although many have hoped, and even hypothesized, that most of these differences could be explained by the fact that boys have elected more advanced mathematics courses than girls have, such has not proved to be true. Even when girls and boys are in the same mathematics classes, starting in middle school, girls, as a group, fall behind boys in important mathematical learning.

One question that should be raised about these sex-related differences in mathematics achievement is whether they appear consistently. The answer is *no*. Sex-related differences in mathematics often do not appear within individual schools. For example, in the Fennema-Sherman (1978) studies, girls were achieving as well as boys in two of the four high schools and in three of the four middle schools that were examined. This reinforces the idea that one should be extremely cautious about generalizing sex-related differences in achievement to an individual. Certainly, many girls achieve at higher levels than many boys and vice versa. Larger variation in achievement is found between high- and low-achieving boys or high- and low-achieving girls (intrasex) than between boys and girls (intersex).

However, the differences that are found are a cause for concern. The major differences are in higher cognitive-level tasks, such as problem solving or application. These types of tasks become more important than computational skills as one progresses in mathematics. It is on these types of tasks that females are falling behind males.

The middle school years appear to be crucial ones in the development of these sex-related differences. Up until that time, few sex-related differences in learning mathematics are found. Near the end of middle school when learners are 13 to 14 years old, differences are often found, with males pulling ahead of females.

Attitudes toward Mathematics

Attitudes influence the learning of mathematics. Most people accept that statement without hesitation, and, indeed, much research shows that specific attitudes and the learning of mathematics are related. (For a thorough discussion of this, see Reyes [1981*a*].) Information about several specific attitudes may help in understanding why females are not achieving as well as males.

Confidence in learning mathematics

Confidence in learning mathematics has to do with the belief that one has the ability to learn mathematics and to perform well on mathematical tasks. It is probably related to, but is different from, enjoying or liking mathematics. It is probably at one end of a continuum, with anxiety being at the other end. The items from the Fennema-Sherman Confidence in Learning Mathematics Scale (Fennema and Sherman 1976) illustrate some facets of this attitude (the Fennema-Sherman Mathematics Attitude Scales are available from the American Psychological Association, 1200 Seventeenth St., NW, Washington, DC 20036):

- Generally, I have felt secure about attempting mathematics.
- I am sure I could do advanced work in mathematics.
- I am sure that I can learn mathematics.
- I think I could handle more difficult mathematics.
- I can get good grades in mathematics.
- I have a lot of self-confidence when it comes to math.
- I'm no good in math.
- I don't think I could do advanced mathematics.
- I'm not the type to do well in math.
- For some reason, even though I study, math seems unusually hard for me.
- Most subjects I can handle OK, but I have a knack for flubbing up math.
- Math has been my worst subject.

Even when girls are achieving as well as boys, they report less confidence in their ability to learn mathematics. This lesser confidence is apparent by at least sixth grade, and maybe earlier. This finding of less confidence by girls lasts throughout high school and undoubtedly influences how hard they study, how much they learn, and their willingness to elect mathematics courses.

Usefulness of mathematics

Another attitude that helps one understand sex-related differences is the perceived usefulness of mathematics. Will it be important in later educational years or in adult life? The items from the Fennema-Sherman Mathematics Usefulness Scale illustrate some facets of students' perceptions of the usefulness of mathematics:

- I'll need mathematics for my future work.
- I study mathematics because I know how useful it is.
- Knowing mathematics will help me earn a living.
- Mathematics is a worthwhile and necessary subject.
- I'll need a firm mastery of mathematics for my future work.
- I will use mathematics in many ways as an adult.

- Mathematics is of no relevance to my life.
- Mathematics will not be important to me in my life's work.
- I see mathematics as a subject I will rarely use in my daily life as an adult.
- Taking mathematics is a waste of time.
- In terms of my adult life, it is not important for me to do well in mathematics in high school.
- I expect to have little use for mathematics when I get out of school.

Starting as early as sixth grade, boys—much more than girls—perceive that mathematics will be useful to them. Although many mathematics educators are reluctant to believe that mathematics is a difficult subject, many learners know it takes hard work to learn it. Why should one work hard to learn mathematics if it has little or no use. Not only does belief in the usefulness of mathematics influence how hard one is willing to work but there is a great deal of evidence that perceived usefulness influences course selection in high school.

Stereotyping mathematics as a male domain

Most people do those things that are seen as appropriate for their own sex. Males try to avoid activities that are perceived as feminine, and females avoid activities stereotyped as masculine. Starting at a very young age, boys, much more than girls, believe that mathematics is a male domain. Examine items from the Fennema-Sherman Math as a Male Domain Scale to understand this attitude:

- Females are as good as males in geometry.
- Studying mathematics is just as appropriate for women as for men.
- I would trust a woman just as much as I would trust a man to figure out important calculations.
- Girls can do just as well as boys in mathematics.
- Males are not naturally better than females in mathematics.
- Women certainly are logical enough to do well in mathematics.
- It's hard to believe a female could be a genius in mathematics.
- When a woman has to solve a math problem, it is feminine to ask a man for help.
- I would have more faith in the answer for a math problem solved by a man than a woman.
- Girls who enjoy studying math are a bit peculiar.
- Mathematics is for men; arithmetic is for women.
- I would expect a woman mathematician to be a masculine type of person.

By sixth grade, boys tend to agree with the last six items and disagree with the first six. This trend continues throughout high school. The finding that boys stereotype mathematics at much higher levels than girls do is pervasive throughout many studies.

How students learn appears to be related to their perception of the appropriateness of what they are studying for their own sex. Girls often learn to read better than boys, and girls do well on high-level cognitive tasks in reading. Reading is stereotyped as being a feminine activity. Boys do better in mathematics, and they view mathematics as a male domain. Certainly, society as a whole tends to present mathematics as a male domain. Most adults who use mathematics a great deal are male. As one goes from elementary to middle to high school the mathematics to be learned increases in difficulty, and the percentage of male mathematics teachers increases. By postsecondary school, most mathematics teachers are male.

Classroom Processes

Suggestions involving processes in the classroom are based on current research. New research on causal attribution of success and failure is explained in *Mathematics Education Research: Implications for the 80's* (Fennema 1981). Although the entire social milieu influences how well one learns mathematics as well as how one feels about mathematics, the most important influences occur within the classroom where mathematics is taught. Learning environments for girls and boys within classrooms, although appearing to be the same, differ a great deal. To start with, teachers interact more with boys than they do with girls. Boys generally receive more criticism for their behavior than girls. Boys often receive more praise and positive feedback than girls. Boys just seem to be more salient in the teacher's view than girls.

A second element of relevant theory deals with autonomous learning. The combination of sex-related differences in achievement and in attitudes that are evident indicates that females, more than males, are not reaching one of the important goals of mathematics education, that of becoming thinkers who are independent problem solvers and who do well in high-level cognitive tasks. Girls, much more than boys, fail to become autonomous in mathematics. Girls develop more negative attitudes related to their ability to perform high-level cognitive tasks—specifically, confidence in learning mathematics and attributional patterns—and this indicates a lack of personal control in mathematics performance.

Many people feel that this differential treatment of girls and boys is partially a result of teachers' differential expectations. The relevant discussion goes something like this: Because of societal beliefs that males are better than females at mathematics, teachers expect that boys will understand high-level mathematics better than girls do and that girls should do better on low-level mathematics tasks like computation. This belief is communicated in a variety of subtle and not-so-subtle ways to both sexes. For example, a teacher might encourage boys more than girls to stick with hard mathematical tasks until solutions are found. The teacher might, with well-intended motivation about preventing failure, assist girls more than boys to find the solution to hard problems. Teachers might call on boys more often to respond to high-level questions and call on girls more often to respond to low-level tasks. Both boys and girls might deduce from these actions that boys are better at high-level tasks and girls are better at low-level tasks. Not only might students conclude that high-level tasks are easier for boys, but they might also conclude that such mathematics is more important for boys, since teachers encourage boys more than girls to succeed in such tasks. Not only would there be these subtle messages, but boys would actually be practicing high-level cognitive tasks more often than girls. Since students learn what they practice, boys would learn to do the problem-solving activities better than girls would.

The hypothesis of differential expectations on the part of teachers is intuitively logical; indeed, Brophy and Good (1974) report that teachers' expectancies are related to the way they interact with students. An interesting study by Becker (1981), done in tenth-grade geometry classes, confirms this also. Becker hypothesizes that the sex-related differences she found in teacher/student interactions were strongly related to differential teacher expectancies. However, other studies designed explicitly to examine teacher expectancies report no differences in expectancy of success in mathematics by teachers for girls and boys (Parsons, Heller, and Kaczala 1980), and students report that teachers have higher expectancies of success for girls than for boys (Fennema et al. 1980). Once again, no clear-cut conclusion can be reached. There appears to be little doubt that teachers, as a group, treat girls and boys (as groups) differently. In other words, the average number of times a group of teachers interact in a specific way with girls and boys differs. However, just as one must avoid the danger of overgeneralizing about individuals and the learning of mathematics, one must also be very cautious about generalizations about how teachers interact with girls and boys. Reyes (1981*b*) presents some interesting findings in this regard. She found tremendous variation in the behavior of twelve seventh-grade teachers.

We do know that when differential interaction patterns occur, the result is evident in the boys' and girls' developing sense of autonomy. Differential behavior results in "learned helplessness" in achievement situations. When students see failure as insurmountable and attribute it to uncontrollable factors, such as a lack of ability, they perceive themselves to be out of control or helpless. In such a situation, they have learned to be helpless and to depend on others for guidance. Dweck (1978) has reported a series of studies that she believes show clearly that teacher behavior directly influences girls to learn to be helpless. Her studies indicate that boys receive criticism from teachers for a variety of reasons relating to both work and discipline whereas girls receive most of their criticism in relation to their work. Dweck believes that because boys receive criticism for both work-related and discipline-related behavior, they learn to ignore all criticism. Because girls receive primarily work-related criticism, they come to believe that any feedback indicates a lack of ability. Learned helplessness is not a well-documented idea but one that is being actively explored. It appears to be worth thinking about, particularly as one teaches mathematics.

Implications for Teaching

When the discussion turns to what teachers ought to do, few hard and fast statements can be made. However, several implications for teaching can be drawn from what we know about sex-related differences in mathematics achievement, attitudes, and participation in classroom processes.

Becoming blind to sex

One general statement appears most reasonable: *Teachers should become blind to sex as they teach mathematics*. Whether a person is female or male might be important in some situations. It is *not* important in the mathematics classroom. What does being sex-blind mean in a classroom? Does it mean ignoring the facts about sex-related differences? No, it means having an even keener awareness of the importance of these facts for all students. In that way, teachers will focus on those aspects of learning mathematics that are important to females in their learning of mathematics. More emphasis on developing feelings of confidence and personal control of the learning of mathematics will help many girls, and many boys as well. Making students—male and female—more aware of the usefulness of mathematics increases the likelihood of more students striving to learn as well as electing mathematics in high school. More instruction in high-level cognitive activities will help all students. Expectations of success should be high for both girls and boys.

Becoming aware of attitudes

One important thing a teacher can do is to become aware of the attitudes that students, both boys and girls, have toward mathematics. Finding this out is not particularly easy, but a number of attitude scales are currently available that have been used with middle school students. (For a complete discussion of this, see Reyes [1981*a*].)

Facing anxieties

Talking about feelings toward mathematics in a nonthreatening way often helps learners put their feelings in a better perspective. The anxieties that students feel are real and will not be alleviated by telling them that it is wrong to have them. However, when teachers get groups of learners to share anxieties, many of them come to realize that such anxieties are not unique and that one is not unusual to have them.

One thing a teacher must be very careful not to do is to lower expectations in the hope of alleviating anxieties. All too often when this happens, what is communicated is that the anxiety has a real basis in fact. A student could easily perceive that when the teacher lowers his or her expectations by, for example, helping in the solution of a difficult problem or giving an easier assignment, the teacher has doubts about the student's ability to do the problem or the assignment. Indeed, many girls report that exactly that has happened to them. They report, for example, that when a sympathetic teacher perceives that they are near tears over a difficult mathematical task, the teacher steps in and does the task. Such action stops the tears but confirms to these girls that they are unable to do the task.

Claiming the credit

Both girls and boys can be taught to be more realistic in their attribution of successes and failures. Hard work or effort usually pays off. Girls, as well as boys, should take the responsibility for their own successes. How often are students encouraged to say, or think, that they succeeded because they were smart or worked hard. When a girl says to the teacher, "Thank you for the A," the teacher could say, "Why thank me? You earned it!" Girls should learn to recognize and accept the idea that they do have ability in mathematics.

Broadening the domain

Teachers should actively combat the belief that mathematics is a male domain. Specific information should be given to girls about the necessity of mathematics for them. Information just as specific should be given to boys about the fact that mathematics is not a male domain. Boys and girls should be led to believe that mathematics is essential for girls as well as for boys. Specific activities should be designed to help boys, as well as girls, understand that mathematics is important for both sexes. We do know that teachers make a difference. Many girls who have succeeded in mathematics talk about teachers who had high expectations or who encouraged them at some point. All too often, girls who have failed remember being discouraged by teachers. Some schools and teachers do what is necessary to see that girls as well as boys succeed. All schools and teachers should.

REFERENCES

Aiken, Lewis R., Jr. "Intellective Variables and Mathematics Achievement: Directions for Research." *Journal of School Psychology* 9 (1971):201–12.

Armstrong, Jane. *Women and Mathematics*. Denver: Education Commission of the States, 1980.

Becker, Joanne Rossi. "Differential Treatment of Females and Males in Mathematics Classes. *Journal for Research in Mathematics Education* 12 (January 1981):40–53.

Brophy, Jere, and Thomas Good. *Teacher-Student Relationships: Causes and Consequences*. New York: Holt, Rinehart & Winston, 1974.

California Assessment Program. *Student Achievement in California Schools*. 1977–1978 Annual Report. Sacramento, Calif.: The Program, 1978.

Dweck, Carol S., William Davidson, Sharon Nelson, and Bradley Enna. "Sex Differences in Learned Helplessness: II. The Contingencies of Evaluative Feedback in the Classroom, and III. An Experimental Analysis." *Developmental Psychology* 14 (1978):268–76.

Ernest, John. *Mathematics and Sex*. Santa Barbara: University of California, 1976.

Fennema, Elizabeth. "The Sex Factor." In *Mathematics Education Research: Implications for the 80's*, edited by Elizabeth Fennema, pp. 92–105. Alexandria, Va.: Association for Supervision and Curriculum Development; Reston, Va.: National Council of Teachers of Mathematics, 1981.

Fennema, Elizabeth, and Thomas P. Carpenter. "Sex-related Differences in Mathematics: Results from National Assessment." *Mathematics Teacher* 74 (October 1981):554–59.

Fennema, Elizabeth, Laurie Hart Reyes, Teri Hoch Perl, Mary Ann Konsin, and Margaret Drakenberg. "Cognitive and Affective Influences on the Developments of Sex-related Differences in Mathematics." Paper read at a symposium of the American Educational Research Association, annual meeting, April 1980, at Madison, Wis.

Fennema, Elizabeth H., and Julia A. Sherman. "Fennema-Sherman Mathematics Attitude Scales: Instruments Designed to Measure Attitudes toward the Learning of Mathematics by Females and Males." *Catalog of Selected Documents in Psychology* 6 (1976):31–32. (*Psychological Abstracts* 56 (1976): No. 1643)

———. "Sex-related Differences in Mathematics Achievement and Related Factors: A Further Study." *Journal for Research in Mathematics Education* 9 (May 1978):189–203.

Glennon, Vincent J., and Leroy G. Callahan. *A Guide to Current Research: Elementary School Mathematics*. Washington, D.C.: Association for Supervision and Curriculum Development, 1968.

National Council of Teachers of Mathematics. *A Position Statement on the Education of Girls and Young Women*. Reston, Va.: The Council, 1980.

Parsons, Jacquelynne E., Kirby A. Heller, Judith L. Meece, and Carol Kaczala. "The Effects of Teachers' Expectancies and Attributions on Students' Expectancies for Success in Mathematics." In *Women's Lives: New Theory, Research, and Policy*, edited by D. McGuigan. Ann Arbor, Mich.: Center for Continuing Education of Women, 1980.

Reyes, Laurie Hart. "Attitudes and Mathematics." In *Selected Issues in Mathematics Education*, edited by Mary Montgomery Lindquist. Chicago: National Society for the Study of Education; Reston, Va.: National Council of Teachers of Mathematics, 1981*a*.

———. "Sex, Confidence in Learning Mathematics, and Classroom Processes." Unpublished doctoral dissertation, University of Wisconsin—Madison, 1981*b*.

Sells, Lucy. "High School Mathematics as the Critical Filter in the Job Market." Unpublished document. Berkeley: University of California, 1973.

Suydam, Marilyn N., and C. A. Riedesel. *Interpretive Study of Research and Development in Elementary School Mathematics*. Vol. 1. Introduction and Summary: What Research Says. Final Report, Project No. 8-0586. Washington, D.C.: U.S. Department of Health, Education and Welfare, 1969.

Suydam, Marilyn, and J. Fred Weaver. *Individualizing Instruction*. Publication of the Interpetive Study of Research and Development in Elementary School Mathematics. Washington, D.C.: Research Utilization Branch, Bureau of Research, U.S. Office of Education, 1970.

3

Help for Learning Disabled Students in the Mainstream

Nancy S. Bley
Carol A. Thornton

THE number of learning disabled students who are being mainstreamed is increasing. It is essential, therefore, that teachers learn and incorporate into their instruction some manageable techniques for dealing with these difficulties in order to effectively plan so that *all* students can learn. A first step toward achieving this goal is to know what is meant by the term *learning disabled.* For the purposes of this article, a learning disabled student is one who has difficulty processing and retaining information due to visual, auditory, language, or motor deficits.

The way these learning difficulties manifest themselves varies greatly from one student to the next. Consequently, what helps one child learn and retain basic mathematical concepts and skills may not help another. Consulting special educators in the school or district is important in this regard. These resource persons can help explain a student's specific learning deficits and suggest general techniques for dealing with them.

Table 3.1 gives examples of the more common learning disabilities that affect the performance of intermediate and upper-grade students in mathematics. These disabilities are grouped under *perceptual, memory,* and *integrative* difficulties and then further divided as to whether they are *visual* or *auditory* deficits. When problems like these are met in the mainstream classroom, teachers need manageable techniques and practices for dealing with them. The students themselves need to be taught ways of personally compensating for, and living with, their handicap. Some suggestions along these lines are contained in the intervention techniques described here. Examples for incorporating each technique in day-to-day instruction in grades 5–8 are also presented. Although techniques like these are helpful to most other students, they are *necessary* for students with specific learning disabilities.

TABLE 3.1

EXAMPLES OF LEARNING DISABILITIES AFFECTING PERFORMANCE IN MATHEMATICS

Type of Learning Disability	Visual Deficit	Auditory Deficit
Perceptual		
Figure-ground	• May not finish all problems on page • Difficulty seeing subtraction within a division problem • Frequently loses place • Difficulty reading multidigit number (see Closure)	• Trouble hearing pattern in skip counting by 5s, 10s • Difficulty attending in the classroom because of extraneous noise
Discrimination	• Difficulty associating operation sign with problem (see Abstract reasoning) • Cannot discriminate among operation symbols • Difficulty reading fractional numbers	• Cannot distinguish between *tenth* and *ten* • Cannot distinguish between *thirty* and *thirteen*
Spatial	• Trouble writing on lined paper • Trouble noticing size differences in shapes • Trouble with fraction concept due to inability to note equal-sized parts • Difficulty writing decimals • Difficulty aligning numbers • Difficulty writing fractional numbers (may also be reversal) • Difficulty with ordinal numbers	• Difficulty following directions using ordinal numbers
Memory		
Short term	• Difficulty copying problems from the board (may be spatial) • Trouble retaining newly presented material	• Difficulty with dictated assignments • Difficulty with oral drills
Long term	• Inability to retain basic facts or processes over a long period	

TABLE 3.1—*Continued*

Type of Learning Disability	Visual Deficit	Auditory Deficit
	• Difficulty solving multi-operation computation	
Sequential	• Difficulty following through on multiplication problems • Difficulty following through on long-division problems • Difficulty solving column-addition problems • Difficulty solving multi-step word problems	• Cannot retain information in dictated word problems
Integrative		
Closure	• Difficulty reading multi-digit number (see Figure-ground) • Difficulty with missing addends and missing factors • Inability to draw conclusions; therefore, trouble noticing and continuing patterns • Trouble continuing counting pattern from within a sequence • Difficulty with word problems	• Difficulty counting on from within a sequence
Receptive language	• Difficulty with words that have multiple meanings • Difficulty relating word to meaning (may be spatial)	• Difficulty writing numbers from dictation • Difficulty relating word to meaning
Expressive language	• Much difficulty with rapid oral drills	• Difficulty explaining why a problem is solved as it is • Difficulty counting on
Abstract reasoning	• Inability to compare numbers using symbols • Inability to solve word problems • Inability to understand patterning in counting • Difficulty with decimal concept	• Inability to follow the logic of oral explanations that focus on the meaning or rationale of a new concept or skill

Intervention Techniques

Model the idea

Children internalize a concept best when they have a mental picture of "what it's all about." At the lower levels, teachers tend to recognize this need more readily. Blocks, play money, and sandpaper numbers are a regular part of the curriculum. These aids are equally as important in the middle and upper grades. The use of fraction rods to illustrate concepts or visual examples at the top of a worksheet are often useful in promoting both understanding and retention. Students with reading, language, or memory deficits will especially benefit from such aids.

Use boxes, circles, and lines

Some children find it difficult to visually sort out what is presented on a textbook page. Perceptually, they confuse the number or diagram they should be focusing on (the *figure*) with surrounding words, symbols, or illustrations (the *background*). Such students are said to have *figure-ground* difficulty.

Boxes. Many children with figure-ground difficulty find it hard to copy problems from a textbook onto paper because visually they do not separate the problems from each other. Box the problems in the mathematics textbook itself so students can readily find them. Separating one problem from another prevents a child from seeing all the problems as one.

Circles. Another common difficulty for learning disabled children having figure-ground deficits is separating the number of the problem from the problem itself. In many books problem numbers are colored to differentiate them from the problems themselves. Even then they may be spatially so close to the problem that children cannot see the difference. Circling the problem number ahead of time has proved to be a very effective way of handling this difficulty.

Plan ahead: boxes, circles, and lines. Before school starts each year, solicit the help of a parent, aide, or other volunteer to prepare textbook pages and worksheets. A template of geometric shapes can be used to help box problems. Problem numbers can also be circled and important directions or examples underlined with bright colors or heavy lines. Eventually children should learn to perform some of these self-help tasks themselves.

Use color coding

Color coding can be an effective way to teach mathematics to many learning disabled children. We have used color effectively to help students learn through their strengths while building up their weaker areas. Research on color coding (Thornton 1981) lends support to the effectiveness of this technique. In particular, we have found that color provides—

1. a way of focusing attention (important for students who have visual perception difficulties or who are highly distractible);
2. a way of properly sequencing steps (important for students with memory or abstract reasoning difficulties);
3. increased ability to recall information;
4. a way of identifying starting and stopping points;
5. a cue to the appropriate response;
6. increased ability to act independently;
7. visual reinforcement for those with auditory difficulties;
8. a way of organizing the visual image to aid those with visual memory difficulties.

Generally it is advisable to code the first step green and the last step red. Children can then be reminded of a traffic light and told to "go on green and stop on red." When more than two colors are used, it is important to choose colors that can easily be distinguished from each other. Do not make step two orange and step three red. Use colored chalk or marking pens during teaching sessions, and later use the same colors on special follow-up worksheets for students who need them. For those few students who are colorblind, heavy or dotted lines can be used instead of colors.

Several examples of the color-coding technique are illustrated in figure 3.1. The first example shows a sample problem completed by a child using colored grid boxes. (Pages containing grid boxes like these can be duplicated in advance and kept on file. Children can be instructed to copy textbook problems onto these pages.) In the example the student first multiplied by the green number (3), and recorded the tens digit in the green circle and the units digit in the appropriate green box. Next the child multiplied 3 times 4 and added the 2 in the green circle. After crossing out the 2, the answer (14) was recorded in the next two green boxes. The student proceeded similarly with the multiplication by 20, this time using the red carrying circle and the red boxes. (Green appears as gray in this example.)

Color can also be used to help students focus on the pattern for placing the decimal point when multiplying, as in example 2, or to help students with abstract reasoning difficulty learn to read decimals by emphasizing the related fraction form, as in example 3. At the symbolic level, fractions are more easily named than decimals. The *10* in the denominator of 1/10 visually cues the child to say *one*-**ten***th*. The *100* in the denominator of 3/100 cues *three*-**hundred***ths*. In decimals, *0.1* or *0.03* of themselves give no similar cues; so an attempt is made to help children associate with the fraction form: "One-tenth has *one* zero in the denominator; tenths (in decimal form) involves *one* decimal place," and so on. This approach has, in our experience, proved quite effective. For those with auditory discrimination deficits, it may also be useful to highlight in color the *th* in decimal

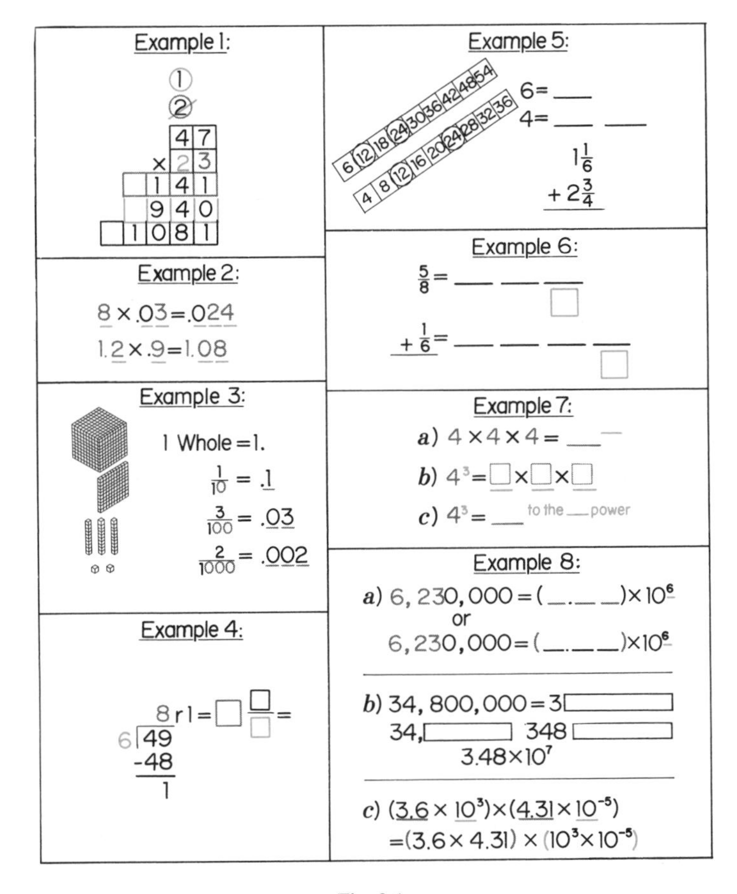

Fig. 3.1

number names to focus attention on the difference between terms like *ten* and *tenth*.

Some students have difficulty writing a quotient as a mixed number. Color-coded pages of problems like the one in example 4 give students the initial practice they need to help them spatially organize their work. Children follow the color cueing and merely copy the numbers in the correct boxes. Then the mixed number is rewritten, independently, for extra reinforcement. Note that the *completed division example* is given at this stage. The focus is specifically to help students learn the correct placement of digits for the mixed-number equivalent of the given quotient. The color scheme controls for extraneous interferences by drawing attention only to relevant digits and their placement. Once students begin to feel more comfortable converting the quotient to a mixed number, have them do the division as well. It is generally a good idea for them first to write the final answer in the form ___ R ___. This method tends to prevent the remainder from being lost in the mass of other numbers. Then the conversion can be made to the mixed-number form.

Example 5 relates to finding the least common denominator in order to add two fractions. At first students may use laminated multiple strips like those shown on which they circle all common multiples. Then they *write out* the multiples of each denominator, stopping with the smallest of the common multiples. Eventually most students will learn to skip count mentally, without the aid of strips, to locate the first (lowest) multiple common to the two denominator numbers. To help students organize their work space when listing the multiples, set up pages as shown. Highlighting in color the space where the lowest common multiple will fall helps children with perseveration or figure-ground deficits know when to stop. Usually, this cue can gradually be withdrawn as children become more sure of themselves. Example 6 shows a simpler system, where the child fills in only the multiple that is marked red. (*Note:* Experience has shown that finding the *least* common multiple, rather than *any* common multiple, results in fewer difficulties, especially for children with expressive language or memory deficits. Numbers are smaller, and there is less need for reducing fractions.)

Exponents can be clarified through color coding. In example 7 students first write the exponential form from a list of factors (7a). Next they practice translating the exponential form (7b). For children with expressive language difficulties it may be necessary to take the sequence a step further (7c).

The idea for scientific notation in example 8a can help students with visual perception or sequencing deficits. An alternative approach using a card is shown in example 8b. When multiplying in scientific notation (8c), students are helped by color highlighting to sort out an otherwise visually confusing field.

Reduce writing tasks

If a student has spatial, motor, or perceptual deficits, it may be necessary to reduce writing tasks. You could assign fewer problems and eliminate or minimize copying from the textbook or chalkboard. Assign, for instance, only every fourth or fifth problem rather than an entire page. Provide worksheets so children don't have to copy problems, making sure there is enough work space on the sheets.

If children are highly distractible, it may also be helpful to—

- create several standard formats for worksheets and provide construction-paper masks to blot out all but a third or a fourth of the page at a time;
- teach students to box off areas on paper into which they can copy problems (this technique helps them organize their work and separate problems from one another);
- keep directions and explanations short and to the point, writing key words to focus attention (this suggestion is also important for children with

auditory memory or auditory perception difficulties);

- actually cut a worksheet into thirds or fourths and assign only *one small section at a time.*

Provide kinesthetic reinforcement

With some children, seeing or hearing is not enough. More total involvement is necessary. The standard procedure to be used parallels that often used in reading: first *finger trace*; next *say;* then *write.* Sometimes children may be instructed to close their eyes while finger tracing a textured pattern. One technique, used for reversals, is finger tracing a textured numeral, then retracing the shape in midair or on paper before writing it.

Another example is that used to reinforce the retention of basic facts. Using the answer side of a flash card, children with visual memory problems might *finger trace both problem and answer.* On turning the card over, students should try to give the answer immediately. If they forget, finger tracing the problem again often triggers the correct response. Sometimes it is helpful to finger trace a problem fact on a child's back. The child is asked to say or write what is felt—problem *and* answer.

A third example emerges as students use region models (fig. 3.2) to find equivalent fractions. Tracing first around the marked part of one region (a) and immediately around the other (b) helps them *feel* the congruence in size. Note that dots have been used rather than conventional shading. This type of marking is more helpful to students with visual perception problems who otherwise have difficulty *finding* the major division lines in the shading.

A final example is based on work with integers. For children who learn kinesthetically, have them spread both arms out and think of their body as a number line. The left arm represents the negative numbers and the right arm, the positive ones. Their body is the zero. To get from positive 3 to negative 5, for instance, they must pass through zero and then on to negative 5. If children do several problems like this, they internalize the feeling and it helps them solve simple addition and subtraction problems without actually using the number line or moving their arms.

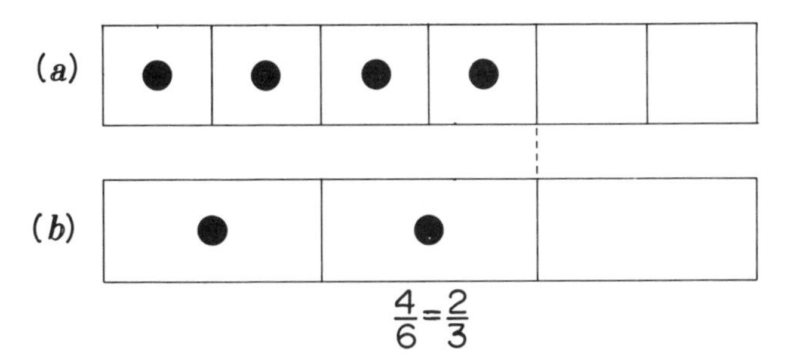

Fig. 3.2

Adapt textbook presentations

For some students, a particular textbook sequence may be inappropriate. For such children you might need to adjust, alter, or reinforce textbook presentations. For example, more and more texts are introducing decimals and decimal computation before computation with fractions. For children with problems in auditory discrimination, spatial organization, or abstract reasoning, it may be necessary to teach naming and writing decimals by directly comparing the decimals to fractions. Toward this end, a more careful review of basic fraction concepts and fraction notation than what most textbooks contain might first be required. As we explained earlier, the written form for fractions is more concrete than that for decimals. It more graphically triggers the visual image of what is represented and hence better cues students on what to say when reading the numeral. Associating the visual model, the fraction, and the equivalent decimal (fig. 3.1, example 3) can help students more readily master otherwise difficult skills for reading and writing decimals.

Another example can be drawn from the area of basic facts. If students have memory difficulties and are familiar only with certain basic facts, it is important to use these facts when presenting a new computational process or topic to the class. This approach eliminates the need for children to draw on weaknesses while trying to learn something new. Controlling the presentation with facts they know allows learning disabled children to learn along with their peers. In the process, known facts are reinforced, an essential aspect of the overlearning that is necessary for mastery. As follow-up, provide a sheet of T's, such as that shown in figure 3.3. Help students black out all facts they have already mastered, and allow them to refer to the sheet whenever they need to. In the meantime challenge them to memorize three or four unknown facts each week.

A final example deals with homework assignments. Instead of assigning all problems from a given page, draw from several pages. Selectively list the problems and the *order* in which they should be completed. Insist that students solve problems in the order specified as a means of developing

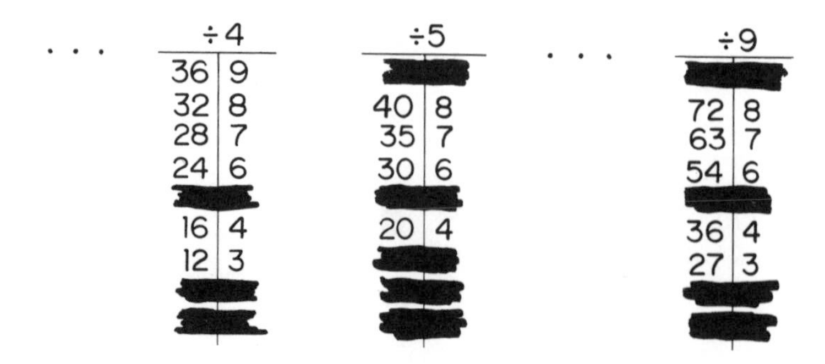

Fig. 3.3

sequencing skills, improving reasoning, and avoiding perseveration. Include a review of previously learned material whenever possible.

Use auditory cueing

Children with visual-perceptual or memory disabilities generally require a high degree of auditory reinforcement. At times, for example, it may be helpful for students to close their eyes to block out distracting visual stimuli and just *listen*. They might read a basic fact and its answer into a tape recorder, then listen to the playback for reinforcement. Many students need to verbalize directions or problems in order to process the information accurately.

Another example relates to naming decimals. Have students work in pairs. One child draws a card like that shown in figure 3.4 and *reads* the decimal. The second student punches the decimal on a hand calculator and checks the display against the card drawn. If necessary, the back of the card can be used to help resolve differences. Alternatively, one child can *read* the decimal on the card while the other uses graph paper strips to picture it. In this and *all areas of mathematics,* it is important that the students be encouraged to express their understanding of important concepts and processes, either verbally or manually, to increase internalization and retention.

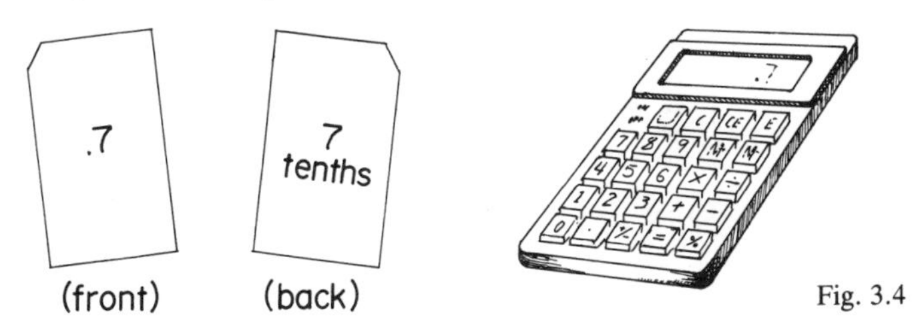

Fig. 3.4

Use patterns or other associations

Many learning disabled children can be helped by instruction that is based on the use of patterns or relevant associations to promote learning and retention. The examples in figure 3.5 illustrate several uses of this technique: relating multiplication by 5 to time on the clock; using size cues to dramatize the comparison of unit fractions; keeping dividends alike and divisors similar within short sequences of practice problems; and providing visual and verbal associations to aid reading larger numbers.

Figure 3.6 focuses on a prerequisite needed for multiplication. The addition skill is developed through coded associations (a) or through the auditory reinforcement of quietly reading the exercise (b). Unless the addition prerequisite is well established, children with sequencing or memory difficulties may have trouble with the multiplication (c).

Provide charts or sample problems

Many classrooms have charts on the walls for cursive letters that stretch all

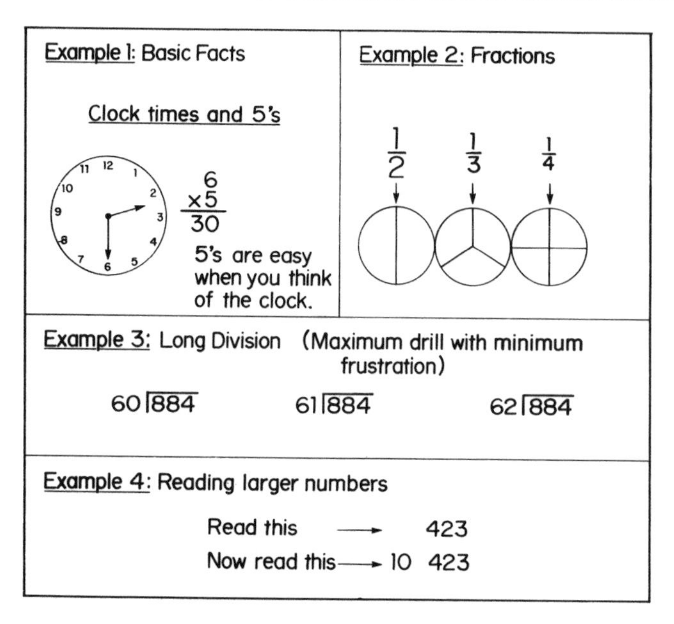

Fig. 3.5

$\begin{array}{r} 4 \\ +4 \\ \hline \end{array}$ so $\begin{array}{r} 54 \\ +4 \\ \hline \end{array}$	$\begin{array}{r} 54 \\ +4 \\ \hline \end{array}$ must end in ___, so 54+4= ___	$\begin{array}{r} 4 \\ 98 \\ \times 6 \\ \hline 588 \end{array}$ 54+4
(*a*)	(*b*)	(*c*)

Fig. 3.6

the way around the room. Be sure to include a number chart as well, even at the upper levels. It will help students who are capable of proceeding mathematically but have reversal or visual-motor difficulties. The chart will serve as an unobtrusive aid to learning disabled students, possibly minimizing their chances of embarrassment because of "not knowing."

Figure 3.7 presents examples of other charts that may be helpful. Example 1 suggests a format for setting up a proportion problem. By filling in the chart with the appropriate information from the original word problem, the student has automatically set up the proportion to solve the problem. Example 2 is helpful to students who have memory sequencing deficits. It also provides the necessary structure for those who cannot provide it for themselves. Since many students enter the fifth or sixth grade with an intuitive understanding of the relationship of percents to fractions and decimals, this structure is especially important. It helps them begin to realize how the conversions are derived. Other charts might list long-division steps (example 3), give rules for computing with integers, or list primes (for reducing

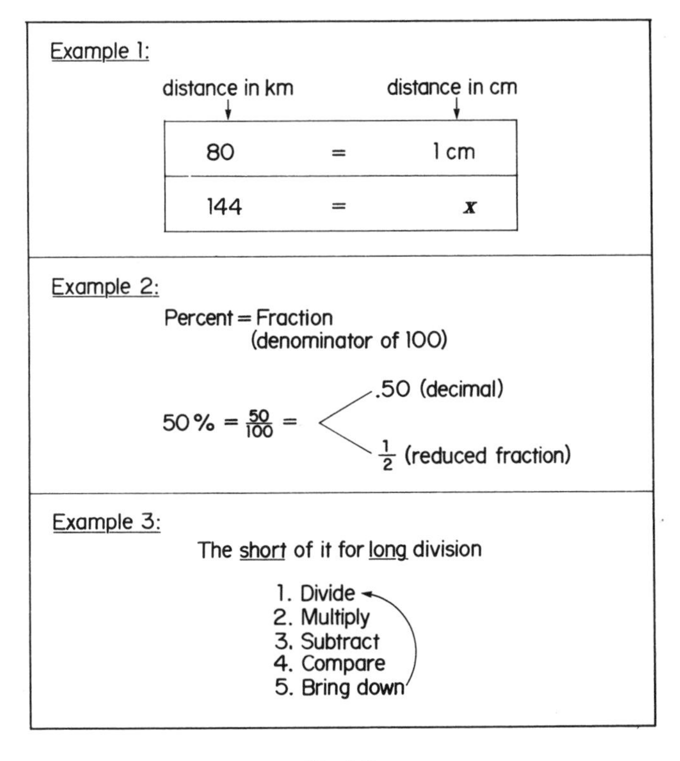

Fig. 3.7

fractions to simple form). Charts like these are necessary references for children in need. Usually they are helpful to the rest of the class as well. An alternative to a wall chart is a personal file card giving the needed information.

To help students who confuse or forget the sequence of a computation, these aids can be used:

- Visual directional clues in a sample problem
- Flip charts for a sample problem, with a separate page for each step
- A sample problem, completed step by step, at the top of a worksheet

Break instruction into segments

For students with learning disabilities, it is critical to sequence instruction in small steps, being careful to provide adequate practice and review. Extra developmental and practice time is necessary for both their understanding and retention of concepts and processes. Breaking instruction into small, meaningful segments makes learning possible rather than overwhelming for these students.

Sometimes it is important to have students complete *just one step* of a process for several examples. For instance, children might be instructed just to place the decimal point in given products or quotients or just to reduce given sums to simple form. Sometimes learning packets with special practice pages can be prepared in advance to support the need for smaller increments and more review. A tape recorder can also be used to provide extra practice for those requiring auditory learning. Students with visually based difficulties are often strong auditorially and can profit from this approach.

Sample Sequence: Writing Fractions

The goal of the following brief sequence is to help students, particularly those with spatial organization, perception, or visual-motor deficiencies, learn to write fractions. It outlines possible activities and exercises that incorporate several of the intervention techniques discussed above.

Once fraction concepts have been established, teachers generally assume that children will have little difficulty actually writing the numbers. This is not always true. We write letters and most numbers in one direction—horizontally, left to right. With fractions, however, the spatial organization is different. Students are now required to write both horizontally *and* vertically. Many learning disabled students cannot clearly discriminate these differences. The sequence of activities and exercises that follows suggests some ways to help children in need write fractions correctly. The major assumption is that difficulties are not due to conceptual misunderstanding.

Instructional sequence

1. **Finger trace.** Felt numbers or a sand or salt tray often help those who need kinesthetic involvement and gross motor activity as prerequisites to writing fractions. Have children trace over a given fraction and then immediately write it again, perhaps with chalk or a large crayon. The gross motor activity helps build the harder, fine motor control needed for standard paper-and-pencil writing of fraction digits with correct placement.

2. **Color code.** When students are ready for paper-and-pencil work, start by using such exercises as that in figure 3.8. The goal is to help children develop the spatial organization, *kinesthetically* and *visually,* to write fractions correctly. As before, the colors are used to help with sequence and number placement. "When you write, green goes first, on top." Coat the green shading and the box outline of several examples with glue. Students can finger trace over the raised surfaces when they are dry. Eventually they

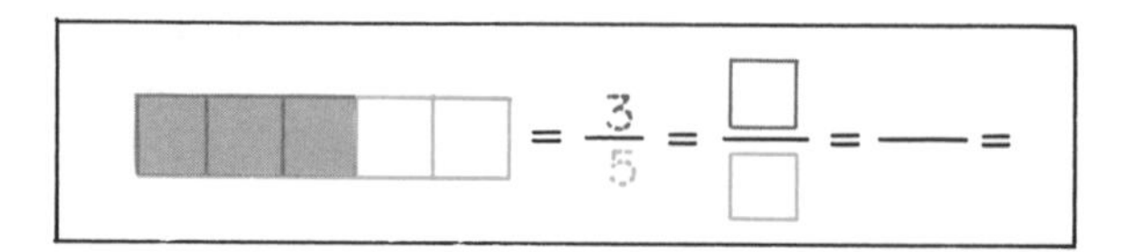

Fig. 3.8

should write the number independently (at the end of the line following the equal sign).

3. **Writing mixed numbers.** Mixed numbers present even greater problems for those with spatial difficulties. Now a sense of midpoint is needed, in addition to a sense of up and down. Figure 3.9 shows one type of exercise that can be prepared in advance and kept on file. The colors are used to help students—

- locate the correct position for each digit;
- develop the correct sequence for writing the numerals;
- associate the parts of the number with the related parts of the picture.

The goal of the assignment is to help students spatially organize their writing of mixed numbers.

4. **Stencil in.** Make stencils available for those students with more severe deficits (fig. 3.10). Have students first use a stencil to write a fraction and then write the same fraction again without the stencil. The two fractions—the stencil and the nonstencil copy—should be written side by side. Then the first serves as a pattern for writing the second.

5. **From words to numerals.** Many students have trouble writing fractions they hear. They don't always place the digits in the correct position. Color-coded exercises (fig. 3.11) often help with both spatial organization and language association. Note that students rewrite the fraction independently as a last step to the exercise.

6. **Fringo.** Fringo, a variation of bingo, can be used to reinforce basic fraction concepts and provide practice in writing fractions. Make a set of cards representing fractions and a set of game boards (all different) with fractions in each square (fig. 3.12). Rather than chips, provide players with plastic overlays and grease pencils. Each player in turn draws a card and states the fraction named. Players who find the symbol on their game board write over it with a grease pencil. The winning pattern should be determined before the game. (*Note:* If necessary, for those with auditory sequencing or

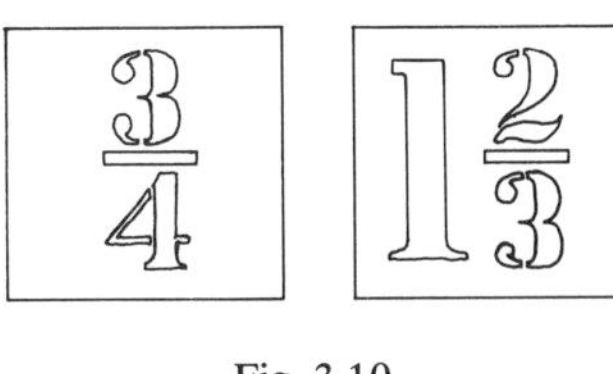

Fig. 3.10

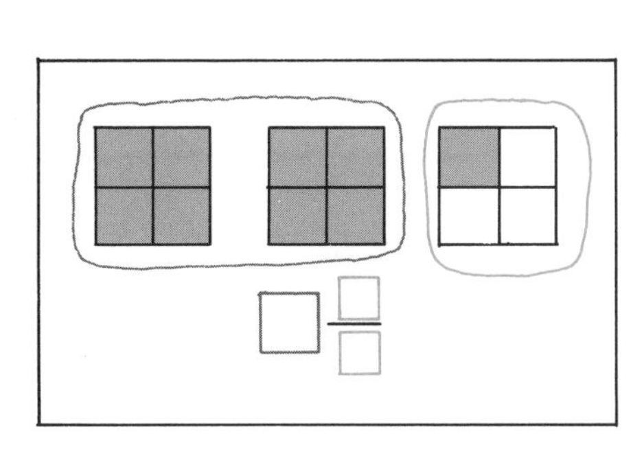

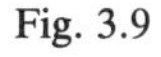

Fig. 3.9

three / fifths = 3 / 5

Fig. 3.11

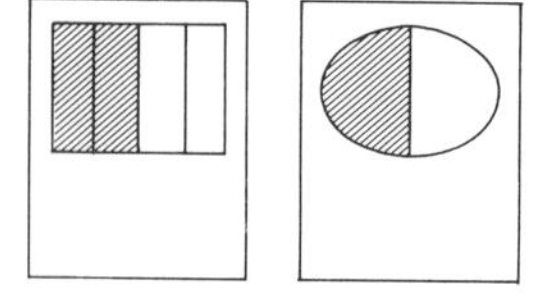

F	R	I	N	G	O
$\frac{1}{2}$	$\frac{3}{4}$	$\frac{2}{3}$	$\frac{1}{7}$	$\frac{5}{8}$	$\frac{1}{3}$
$\frac{3}{8}$	$\frac{5}{10}$	$\frac{3}{4}$	$\frac{3}{10}$	$\frac{4}{5}$	$\frac{3}{7}$
$\frac{7}{10}$	$\frac{1}{2}$	$\frac{2}{3}$	$\frac{7}{8}$	$\frac{5}{6}$	$\frac{6}{8}$
$\frac{9}{10}$	$\frac{1}{5}$	$\frac{1}{8}$	$\frac{1}{10}$	$\frac{3}{8}$	$\frac{5}{6}$
$\frac{3}{4}$	$\frac{1}{2}$	$\frac{1}{6}$	$\frac{5}{8}$	$\frac{7}{8}$	$\frac{3}{6}$

Fig. 3.12

memory problems, the cards can be color coded. Use the suggestions from the exercises of this section for the coloring scheme. Throughout these exercises, bold, dotted, or regular line drawings can be used instead of color coding for color-blind students.)

Discussion

The intervention techniques outlined in this article have proved most helpful in our own work with learning disabled students. We have tried and tested them both in regular classrooms and in clinical settings. These techniques are important and necessary approaches for teaching mathematics to students with specific learning difficulties; they are helpful for most other students as well.

BIBLIOGRAPHY

Bley, Nancy S., and Carol A. Thornton. *Teaching Mathematics to the Learning Disabled.* Rockville, Md.: Aspen Systems Corp., 1981.

Cawley, John F., et al. "Mathematics and Learning Disabled Youth: The Upper Levels." *Learning Disability Quarterly* (Fall 1978): 37–52.

Flinter, Paul F. "Educational Implications of Dyscalculia." *Arithmetic Teacher* 26 (March 1979): 42–46.

Herold, Persis Joan. "Reasons for Failure." *Exceptional Parent* 9 (February 1979): 49–53.

Houck, Cherry, Robert M. Todd, Doris H. Barnes, and Judy B. Englehard. "LD and Math: Is It the Math or the Child?" *Academic Therapy* 15 (May 1980): 557–70.

Reisman, Fredricka K., and John F. Riley. "Teaching Mathematics to LD Adolescents." *Focus* 1 (October 1979): 67–73.

Steeves, Joyce. "My Math Is All Right, What's Wrong Is My Answers." *G/C/T* 12 (March-April 1980): 52–57. (G/C/T, Box 55564, Mobile, AL 36606)

Thornton, Carol A. "The Effects of Color Cueing for Promoting Learning of Selected Fractional Topics by Slow Learners and Learning Disabled Students." Unpublished research, supported by Illinois State University through its Faculty Research Grant Program, 1981.

Thornton, Carol, Edna Bazik, John Dossey, and Benny Tucker. *Teaching Mathematics to Children with Special Needs.* Menlo Park, Calif.: Addison-Wesley Publishing Co., 1981.

Thornton, Carol A., and Rosemary Reuille. "The Classroom Teacher, the LD Child, and Math." *Academic Therapy* 14 (September 1979): 15–21.

4

Integrating Computer Literacy into the Curriculum

Carla J. Thompson

COMPUTERS are rapidly becoming as commonplace as television in our society. To fully prepare students to function in a computer-oriented society, education must foster computer literacy. Computer literacy is defined as the ability to exchange information "with" or "about" computers. In terms of the student learner in mathematics, computer literacy involves (1) working directly *with* computers in practice, programming, and simulation exercises and (2) developing an understanding *about* the role of computers in society, the relationship of mathematics and computers, and the interrelationship of mathematics and computer literacy in career planning. For students to participate effectively in a society that depends heavily on computerized information, computer literacy must be as much an educational obligation as reading literacy (Molnar 1978–79).

Already middle school students are beginning to become aware of the feasibility of a computer in every home. Because of the increasing availability and ever-lowering cost of microcomputers and the continual advancements in technology, the responsibility for preparing students for a changing society becomes greater each day. During the late 1970s and early 1980s, writers concerned with the school's responsibility for developing a computer-literate society urged educators in all disciplines to introduce computing into the curriculum (Dennis 1978; Dwyer 1971; Milner 1979; Molnar 1978–79; Poirot 1979; Taylor et al. 1979). Mathematics education must assume an increasing role in this process. Mathematics and its applications in problem solving are the basis for our technological future. Middle school students must receive an adequate computer education to function in society.

Efforts to define computer literacy for curriculum planning prompted writers to begin identifying specific computing skills and competencies that were needed in computer literacy programs (Johnson et al. 1980; Klassen 1978; Wolmut et al. 1979). Mathematics educators developed lists of cognitive and affective computer literacy objectives for computer literacy courses (Johnson et al. 1980; Koetke 1978; Wolmut et al. 1979).

Although such lists offered a much-needed organization and specification of objectives for computer literacy courses, the interjection of a computer literacy course into the middle school curriculum did not seem feasible. The middle school curriculum is already overcrowded. Compounding this problem, middle school teachers unfamiliar with computer literacy skills and lacking in computer competencies may not have the ability or confidence to include computer literacy objectives and activities in their teaching. Thus, a plan is needed for integrating computer literacy training into the existing curriculum.

The Middle School Computer Literacy Mathematics Curriculum Integration Matrix (fig. 4.1) was developed in an attempt to satisfy this need. Content areas that are representative of the mathematics units covered in grades 5 to 9 were selected as the horizontal basis for the matrix. Computer literacy areas were identified from the literature (Johnson et al. 1980; Koetke 1978; Wolmut et al. 1979) and grouped according to their direct computer influence or their computer awareness; these areas formed the vertical basis for the matrix. The matrix contains major objectives (indicated by shading) and minor objectives as instructional suggestions for integrating computer literacy with mathematics.

Lists of illustrative major and minor objectives and activities are given below for each mathematics unit represented in the matrix. Major objectives and activities are suggested as necessary instructional objectives for integrating computer literacy, and minor ones are offered to be used at the teacher's discretion; for instance, because of time constraints the teacher may wish to ignore some or all minor objectives and activities or use them as activities for student enrichment.

Computer literacy areas classified as direct computer influences contain objectives and activities that require the availability of hardware (calculators, computer terminals or printers, and graphics equipment), software (commercial and student-written programs), and some background in computer science for the teacher. The computer-awareness objectives and activities, however, require no computing equipment in the classroom and no previous knowledge in computer science for the teacher.

The integrated approach presented here offers a plan by which middle school students can develop computer literacy skills within the existing mathematics curriculum. Direct computer literacy areas suggested by the matrix promote student involvement with computers in the classroom. Middle school teachers who lack computer literacy skills and computing competencies can find in the computer awareness areas some instructional activities that do not depend on previous knowledge in computer science.

Although the objectives and activities that are suggested give teachers a basis for planning and making decisions, the matrix only begins to tap the list of possible computer literacy skills that can be integrated into the middle

MIDDLE SCHOOL COMPUTER LITERACY
MATHEMATICS CURRICULUM INTEGRATION MATRIX

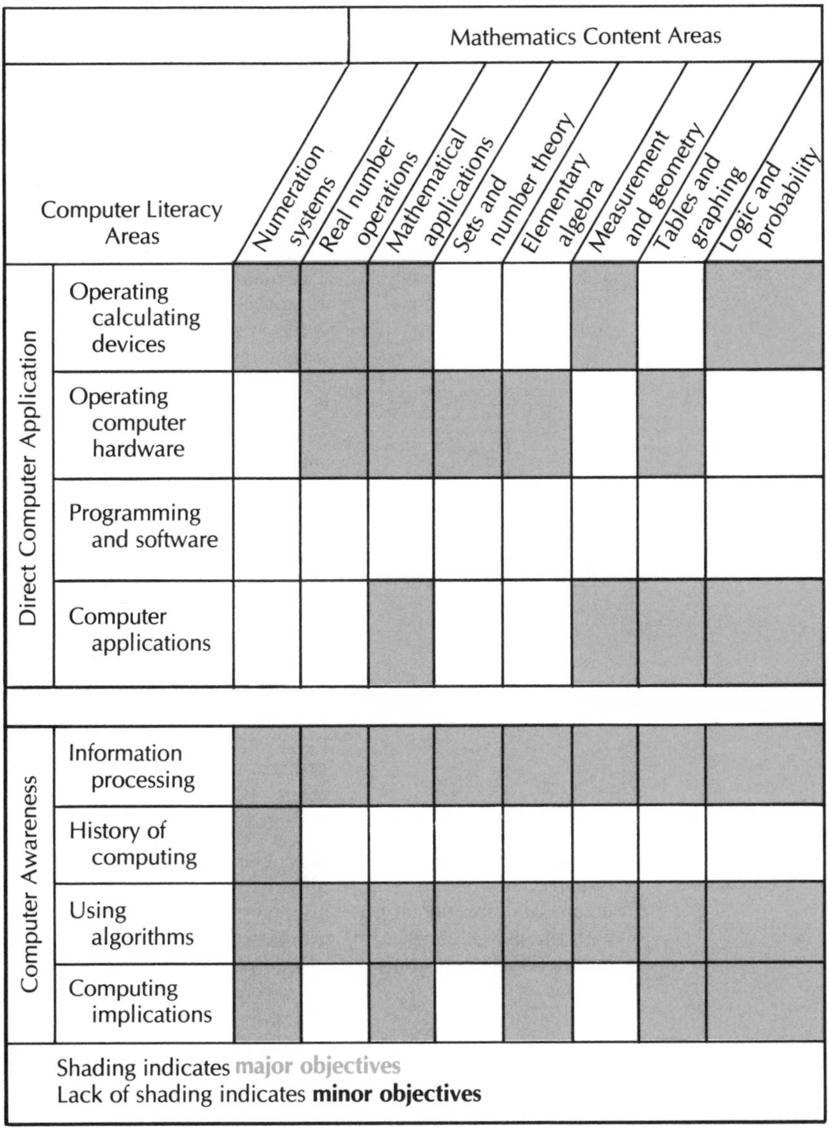

Computer Literacy Areas		Numeration systems	Real number operations	Mathematical applications	Sets and number theory	Elementary algebra	Measurement and geometry	Tables and graphing	Logic and probability
		Mathematics Content Areas							
Direct Computer Application	Operating calculating devices	■	■	■			■		■
	Operating computer hardware		■	■	■	■		■	
	Programming and software								
	Computer applications			■			■	■	■
Computer Awareness	Information processing	■	■	■	■	■	■	■	■
	History of computing	■							
	Using algorithms	■	■	■	■	■	■	■	■
	Computing implications	■		■		■		■	■

Shading indicates major objectives
Lack of shading indicates **minor objectives**

Fig. 4.1. Matrix for integrating computer literacy into the middle school curriculum

school mathematics curriculum. However, time and resource constraints make complete coverage of computing competencies impractical. Therefore, selecting and implementing appropriate computer literacy skills is subject to the teacher's consideration and planning. Ultimately, the effective integration of computer literacy into the middle school mathematics curriculum depends on the teacher.

UNIT: NUMERATION SYSTEMS

Computer Literacy Areas	Specific Objectives (Student Directives)	Classroom Applications (Teacher Directives)
Operating calculating devices	Demonstrates place value using various types of nonelectronic calculating devices	Encourages pairs of students to participate in hands-on counting and grouping activities using abaci, Cuisenaire rods, and counters
Operating computer hardware	Uses drill and practice programs related to numeration systems	Acquaints students with the procedure of accessing a file at a keyboard terminal by having one student demonstrate a program that requires students to practice changing base-two numbers to base ten
Programming and software	Compares characteristics of programming or coding with characteristics of numeration systems	After introducing students to the process of writing a simple BASIC program, asks students to generate a list of organizational characteristics that BASIC programming and the base-ten numeration system have in common (e.g., placement of statements in programming and placement of digits in writing numerals)
Computer applications	Compares the use of numeration systems as a means of communication with the use of computers in telecommunications	Discusses with students the universal nature of numeration systems in communicating numerical information and the universal nature of computer applications in telecommunications (e.g., international communications by telephone and satellite, international airline reservations)
Information processing	Recognizes that information in digital computers is stored as electrical representations of zeros and ones	Gives students several drawings of light panels with various ON/OFF states and asks them to decode the panels in both binary and decimal representations

Computer Literacy Areas	Specific Objectives (Student Directives)	Classroom Applications (Teacher Directives)
History of computing	Discusses characteristics of early calculating devices	Asks students to research and report to the class about the origin and development of early calculating devices, such as the abacus, Napier's bones, and so forth
Using algorithms	Flowcharts procedures for changing numbers from one numeration system to another	Assigns groups of students various numeration systems (base five, base two, etc.) to change to base ten by flowcharting the appropriate procedures; then has the entire class develop a general algorithm for converting any base to base ten
Computing implications	Discusses technological implications of the binary system, for example, operational logic of circuits	Using slides or films, presents students with examples of the use of electronic circuits in different types of technical machinery
UNIT: REAL NUMBER OPERATIONS		
Operating calculating devices	Recognizes that checking results for correctness is part of the problem-solving process in computing	Encourages pairs of students to check one another's work using hand-held or desk-top calculators
Operating computer hardware	Accesses and uses drill-and-practice programs that involve operations with real numbers	Uses commercial drill-and-practice programs involving real number operations or has students write their own drill-and-practice programs
Programming and software	Develops and codes programs that will generate simple addition, subtraction, multiplication, and division problems for drill	Introduces students to appropriate BASIC coding for the operations of addition, subtraction, multiplication, and division, including the random number generator function
Computer applications	Recognizes that a major use of the computer is	Invites a weather forecaster to visit the class as a

Computer Literacy Areas	Specific Objectives (Student Directives)	Classroom Applications (Teacher Directives)
	for numerical applications, such as weather prediction	guest speaker, or plans a field trip to a local weather station
Information processing	Organizes procedures for working with real numbers by analyzing order of operations	Gives students numerical computation problems involving more than one arithmetic operation and asks students to list the appropriate order of operations
History of computing	Explains contributions of major figures (Napier, Pascal, Babbage) in the historical development of mechanical computation	Gives students time to research and write reports on the major contributions of Napier, Pascal, Babbage, and others, and to present these reports to the class
Using algorithms	Flowcharts problems involving one or more arithmetic operations	Instructs groups of students to create, with long strips of butcher paper and markers, flowcharts using arithmetic operations; then asks groups to exchange flowcharts for solving
Computing implications	Recognizes limitations in relying on calculators rather than on pencil and paper for arithmetic computations (e.g., mechanical failure, limited number of processes available, and limited accuracy)	Gives students several types of computational problems, such as fractions, decimals, and percents, that involve floating decimals and rounding; then asks students to list some possible limitations these problems might create for calculators
UNIT: MATHEMATICAL APPLICATIONS		
Operating calculating devices	Uses desk-top or hand-held calculators in computing interest	Has students participate in a role-playing situation involving customers desiring to procure bank loans, including completing loan applications, discussing interest rates and collateral with the loan officer, cal-

Computer Literacy Areas	Specific Objectives (Student Directives)	Classroom Applications (Teacher Directives)
		culating interest, and determining terms for the loan
Operating computer hardware	Accesses and uses a computer program for balancing a checkbook	Using student-made checkbooks for input data (construction paper and imagination work well together), encourages individual students to go through the process of balancing a complete group of deposits and withdrawals using a computer terminal and an appropriate computer program
Programming and software	Distinguishes between truncation (the process of "chopping" digits) and rounding in computer application problems	Allows pairs of students to examine computer printouts of money application problems (e.g., computing sales tax or investment problems) to understand the computer processes of truncation and rounding
Computer applications	Describes societal institutions that are increasingly dependent on computers	Plans several field trips to banks, department stores, grocery stores, and large companies that use computers in mathematical problem-solving situations
Information processing	Translates word problems into equivalent mathematical equations	Gives students lists of words often found in word problems and asks them to supply the equivalent mathematical terms
History of computing	Discusses technological improvements that have occurred in the development of the computer since its inception (e.g., vacuum tube, transistor, integrated circuit) and reasons for the rapid advancement in the computer field	Requests a visit from a guest speaker who works in the electronics area, especially someone who works with microprocessors, to discuss technological improvements in the development of the computer

Computer Literacy Areas	Specific Objectives (Student Directives)	Classroom Applications (Teacher Directives)
Using algorithms	Flowcharts the procedures involved in balancing a checkbook	For a class project, divides the class into two groups (deposits and withdrawals) and asks each group to flowchart their procedures and then combine the flowcharts into a class-constructed flowchart
Computing implications	Recognizes technological developments that have reduced costs and made computers accessible for personal use in society	Plans a field trip for students to attend a microcomputer exhibition; if this is not possible, requests a demonstration from a microcomputer vendor
UNIT: SETS AND NUMBER THEORY		
Operating calculating devices	Uses calculators to develop series of numbers	Encourages pairs of students to complete arithmetic and geometric series using calculators to compute sequential values
Operating computer	Assesses and runs a program that will sort a given set of numbers into increasing or decreasing order	Allows pairs of students at keyboard terminals to use bubble sort and ripple sort programs (preferably student-written programs) to sort a set of numbers; encourages class discussion of differences between the two sorting procedures
Programming and software	Writes a simple BASIC program to sort a set of numbers	Introduces students to computer sorting procedures, such as bubble sort or ripple sort, used in ordering sets of numbers in increasing or decreasing order
Computer applications	Describes various uses of the sorting capability of the computer in business and industry	Uses films to depict record-keeping uses of computers (available from large computer companies); examples include locating parts, accessing al-

Computer Literacy Areas	Specific Objectives (Student Directives)	Classroom Applications (Teacher Directives)
		phabetical records, ordering post office ZIP codes, and so forth
Information processing	Works with elements of set theory that apply to computer science (e.g., subsets, Venn diagrams, union, intersection, and empty set)	Gives students various written exercises involving operations with sets, especially work with *and* and *or* (union and intersection)
History of computing	Describes historical advancements of factors, such as memory and speed, that relate to efficiency in computer sorting techniques	Invites a guest speaker familiar with computer systems design to discuss the advancements in computer memory and speed
Using algorithms	Creates a flowchart for generating number sequences and series	Emphasizes the use of looping in flowcharting; encourages students to use loops to generate number series
Computing implications	Discusses examples of searching and sorting everyday objects	Asks class as a group to compile a list of resources used in everyday life that illustrate searching and sorting procedures (e.g., an index, a dictionary, a phone book, the card catalog in the library, etc.)
UNIT: ELEMENTARY ALGEBRA		
Operating calculating devices	Compares the speed and accuracy of various calculating devices	Has students solve simple linear equations (involving only one or two arithmetic operations) using first a slide rule and then a calculator for the necessary computations
Operating computer hardware	Accesses and uses a computer program for practice in finding solutions to open sentences	Asks individual students to use a drill-and-practice program for finding solutions to open sentences
Programming and software	Recognizes that computers can be programmed for solving linear equations (linear programming)	Introduces appropriate BASIC coding used to indicate variables and arrays

Computer Literacy Areas	Specific Objectives (Student Directives)	Classroom Applications (Teacher Directives)
Computer applications	Discusses scientific and mathematical applications of computers	Invites several guest speakers (e.g., electrical engineer, research mathematician, geophysicist, or nuclear physicist) to visit the class and discuss various uses of computers in their work
Information processing	Translates word problems into equivalent mathematical equations	After students have worked some exercises with word problems, asks them to use their solutions and write one or two sentences that will satisfactorily answer the questions posed by the problems
History of computing	Compares various methods that have been used through the years for putting information (numeric and alphanumeric) into computers	Allows students to examine various forms of input materials that have been used through the years to represent information (e.g., punched cards require two holes to represent variables or letters and one hole for numbers)
Using algorithms	Flowcharts algebraic processes involving polynomials	Selects groups of students to develop flowcharts depicting appropriate procedures for adding, subtracting, multiplying, and dividing polynomials
Computing implications	Describes the appropriate mathematics background for various computer-related occupations	Allows students to select for investigation one computing occupation (such as computer operator, programmer, systems analyst, etc.) and to report to the class; has students include the responsibilities in each career and the necessary mathematics preparation for each

Computer Literacy Areas	Specific Objectives (Student Directives)	Classroom Applications (Teacher Directives)
UNIT: MEASUREMENT AND GEOMETRY		
Operating calculating devices	Uses a calculator to compute surface area and volume problems	Has students measure several solid objects and calculate their surface area and volume to encourage the use of several mathematical instruments in solving.
Operating computer hardware	Accesses and uses computer game programs involving geometric situations	Investigates commercial game and simulation programs such as spatial relationships and geometric pattern strategies
Programming and software	Writes a program in BASIC that will generate perimeter and area problems	Shows students how to combine the integer and random number functions in BASIC to ensure that only positive integer values will be generated as dimensions for perimeter and area problems
Computer applications	Recognizes measurement and geometric applications of the computer in industrial and technical fields, such as machine technology, construction, welding, and so forth	Discusses with students several examples of measurement and geometric applications of computers, such as converting English measurements to metric, producing appropriate angle relationships, and scaling proportions in blueprint reading
Information processing	Determines equivalent measures within measurement systems	Asks students to practice finding equivalent measures within the metric system (e.g., changing meters to centimeters)
History of computing	Compares the evolvement of measuring instruments with the development of calculating and computing devices	For a class project asks students to create two historical time lines entitled "History of Measurement" and "History of Computing" on two long strips of butcher

Computer Literacy Areas	Specific Objectives (Student Directives)	Classroom Applications (Teacher Directives)
		paper so that drawings, pictures, and written descriptions can effectively depict the stages in the development of measurement and computing instruments
Using algorithms	Flowcharts procedures involved in geometric constructions	Assigns each student one type of geometric construction to flowchart (such as constructing perpendicular lines, parallel lines, bisecting an angle, etc.); then, to check the logic and accuracy of the procedures, asks students to exchange flowcharts and perform the constructions indicated using the step-by-step procedure
Computing	Describes present and future measures that depict computer access and processing time	Discusses with students computer measurements of time (such as the microsecond and nanosecond) for accessing and processing data; discusses possible measures of the future
UNIT: TABLES AND GRAPHING		
Operating calculating devices	Uses calculators to calculate percentages or other computations to complete graphs or supply information for tables	Gives students partially completed tables of information and has them fill in the appropriate information
Operating computer hardware	Recognizes components of computer hardware that support graphics by observing demonstrations or operating graphic equipment	Calls in a computer vendor, if possible, to demonstrate graphic capabilities and such equipment as plotters, joysticks, light pens, and so forth
Programming and software	Writes a simple program to produce a table of information as output	Introduces students to spacing, formatting, and tab functions used in programming to produce tables as output

Computer Literacy Areas	Specific Objectives (Student Directives)	Classroom Applications (Teacher Directives)
Computer applications	Discusses examples of computer-enhanced pictorial displays, such as geophysical pictures, space photography, and X rays, and their business and science uses	Has students bring pictures of examples of computer graphics used in business, the space industry, medical applications, and so forth, for a bulletin-board display of computer graphics
Information processing	Decodes and interprets information from charts, tables, and pictorial and graphic representations	Asks students specific questions to pinpoint information contained in bar graphs, line graphs, or other graphic representations
History of computing	Recognizes technological developments that have improved communication between humans and computers, such as graphics or pictorial representations and voice synthesis	Obtains one of the films that are available from the larger computer corporations to demonstrate graphic and vocal capabilities of computers
Using algorithms	Plans a step-by-step procedure to determine the appropriate format and organization of information for drawing tables or graphs (bar graphs, line graphs, circle graphs, etc.)	Has students draw flowcharts depicting steps involved in constructing tables or graphs; includes such decision questions as which type of graph is most appropriate, how will information be organized, and so forth
Computing implications	Recognizes that the graphic capability of computers affords the opportunity to view large quantities of information from different perspectives quickly and accurately	Supplies two teams of students with demographic information, such as data from the U.S. census—one team receives a list of numerical data and the other, a computer printout of the same data; asks both teams to answer specific questions to demonstrate the efficiency of using the computer printout

Computer Literacy Areas	Specific Objectives (Student Directives)	Classroom Applications (Teacher Directives)
UNIT: LOGIC AND PROBABILITY		
Operating calculating devices	Describes the probability of an event in terms of randomness and experimental results	Encourages groups of students to conduct experiments involving drawing cards, tossing coins, or rolling dice and then to use calculators to compute probabilities
Operating computer hardware	Operates a simulation probability program involving flipping coins, rolling dice, or drawing cards	Obtains commercial programs for students to practice probability simulations
Programming and software	Identifies logical inconsistencies in the coding of a program	Has groups of students read, interpret, and analyze the coding of a computer program, line by line, for inconsistencies in logic—for example, GO TO and IF THEN procedures
Computer applications	Recognizes the use of computers in business and industrial projection studies	Asks students to examine graphic examples of computer production or profit projection graphs
Information processing	Identifies logical difficulties and inconsistencies	Allows students to experiment with logic exercises that involve contradictions, inappropriate inferences, and incomplete sequences
History of computing	Makes conjectures about futuristic applications	Encourages students to read and report to the class various scientists' futuristic views of the uses of computers—such as robots—in logical decision making
Using algorithms	Distinguishes between logical and illogical inferences in cause-and-effect situations	Asks students to analyze paragraphs involving cause-and-effect situations by drawing out and listing

Computer Literacy Areas	Specific Objectives (Student Directives)	Classroom Applications (Teacher Directives)
		the step-by-step logical statements
Computing implications	Recognizes assertions or conclusions that may be unsupported by accompanying data or distorted by personal or institutional bias	Asks students to collect newspaper or magazine advertisements or to copy TV commercials and have groups of students examine them for logical, accurate reporting versus biased or inconsistent information

REFERENCES

Dennis, J. Richard. "Undergraduate Programs to Increase Instructional Computing in Schools." In *Proceedings of the Ninth Conference on Computers in the Undergraduate Curricula,* edited by Ronald E. Prather, pp. 212–18. Denver: University of Denver, 1978.

Dwyer, Thomas A. "Some Principles for the Human Use of Computers in Education." *International Journal of Man-Machine Studies* 3 (1971): 219–39.

Johnson, David C., Ronald E. Anderson, Thomas P. Hansen, and Daniel L. Klassen. "Computer Literacy—What Is It?" *Mathematics Teacher* 73 (February 1980): 91–96.

Klassen, Daniel L. "Introduction to Computers in the Curriculum." In *Computer Applications in Instruction*, edited by Diana Harris and Laurie Nelson-Heem, Northwest Regional Educational Laboratory, pp. 109–11. Hanover, N.H.: Time Share Corp., 1978.

Koetke, Walter. "Computers in the Mathematics Curriculum." In *Computer Applications in Instruction,* edited by Diana Harris and Laurie Nelson-Heem, Northwest Regional Educational Laboratory. Hanover, N.H.: Time Share Corp., 1978.

Milner, Stuart P. "An Analysis of Computer Education Needs for K–12 Teachers." In *Proceedings of the NECC 1979 National Educational Computing Conference,* pp. 27–30. Iowa City: University of Iowa, 1979.

Molnar, Andrew R. "The Next Great Crisis in American Education: Computer Literacy." *Journal of Educational Technology Systems* 7 (1978–79): 275–83.

Poirot, James L., John Hamblen, James D. Powell, and Robert P. Taylor. "Computing Competencies for School Teachers: A Preliminary Projection for the Teacher of Computing." In *Proceedings of the NECC 1979 National Educational Computing Conference,* pp. 36–38. Iowa City: University of Iowa, 1979.

Taylor, Robert P., James L. Poirot, James D. Powell, and John Hamblen. "Computing Competencies for School Teachers: A Preliminary Projection for All But the Teacher of Computing." In *Proceedings of the NECC 1979 National Educational Computing Conference,* pp. 39–43. Iowa City: University of Iowa, 1979.

Wolmut, Peter, Jim Ylvisaker, and Bob Allenbrand. *K–12 Course Goals in Computer Education.* Project of the Oregon Local and Education Service Districts of Clackamas, Multnomah, and Washington Counties. Portland, Oreg.: Commercial-Educational Distributing Services, 1979.

5

Developing Mental Arithmetic

Bill Atweh

A GROUP of elementary school teachers was given the following problem to solve:

$$\begin{array}{r} 281 \\ +\underline{170} \end{array}$$

Although no instructions were given, none of those present felt a need to use paper and pencil. Most of the teachers came up with the correct answer of 451. The teachers were then asked to explain the reasoning they followed to obtain the answer. Although their methods revealed some individual differences, the following reasoning was typical:

$$\begin{array}{rcl} 200 + 100 &=& 300 \\ 80 + 70 &=& 150 \\ 300 + 150 &=& 450 \\ 450 + 1 &=& 451 \end{array}$$

Not a single participant started by adding the units digits. When the teachers were asked whether this was the way they taught addition to their classes, many were taken aback. The question "Why not?" was met by complete silence.

A second group of teachers was given the following problem:

> You go to a store to buy four tapes. Each tape costs $2.35. You have $10 with you. Can you pay for all four tapes?

This problem required a little more time to solve than the first one. However, most teachers came up with a positive answer to the question. Again, when they were asked to explain their reasoning, some gave the following:

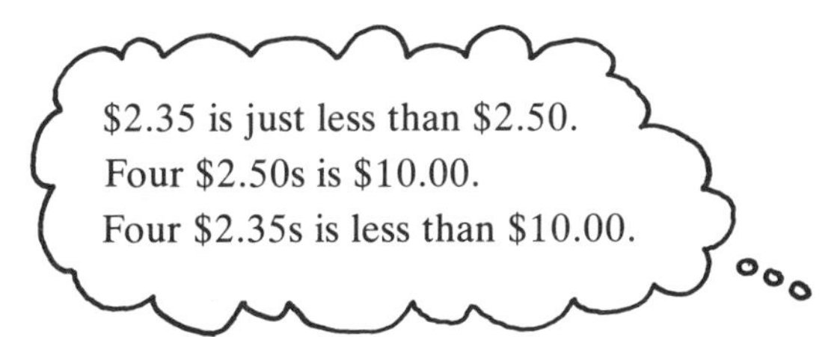

An experienced primary school teacher argued that the repeated subtraction method to find quotients should never be taught in the schools. "It is archaic, cumbersome, and eventually useless," he added. He was given the following problem:

How many 23s in 276?

Not having access to paper and pencil or a calculator, the teacher reasoned:

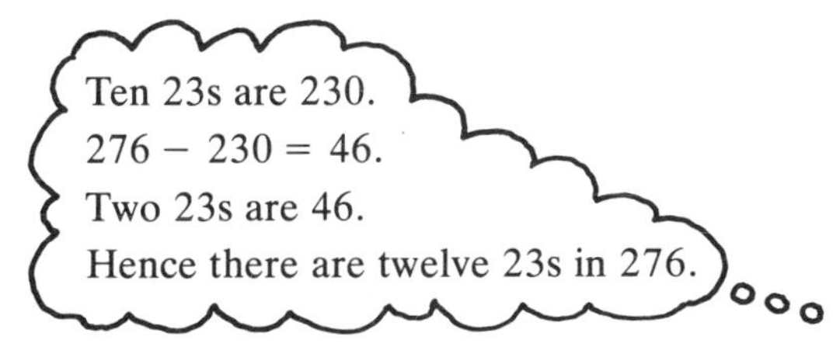

These three problem situations illustrate some of the procedures that people use in solving problems or performing computations without the assistance of paper and pencil, a calculator, or any other concrete aid. All three methods are different from (some are directly opposite to) those algorithms normally used in school mathematics. They are used either because they are simpler or because they require less strain on the memory than the "usual" algorithms. This category of techniques is often called mental arithmetic.

Mental Arithmetic and Estimation

Many teachers think of the terms *mental arithmetic* and *estimation* as synonyms. Yet some educators have dealt with the two concepts independently of each other. Although the terms are strongly related in the practice of teaching mathematics, it is important to realize that they stand for distinct concepts. Mental arithmetic is a method of thinking through a problem, performing an operation, or obtaining a result, as opposed to using paper and pencil or some other concrete aid. Estimation refers to one aspect of the result obtained, namely, its precision, as opposed to an exact answer or solution. Since the method used and the precision of the answer are two independent dimensions of computation, they can be paired together to yield four different types of response activities in which children can be involved (fig. 5.1).

		Method	
		Concrete	Mental
Precision	Exact	1	2
	Estimate	3	4

Fig. 5.1

Four students are given the problem 271.25 + 130.59. Each reasons and responds in a different way, and each of these ways can be placed in one of the categories in figure 5.1.

Category 1. Robert writes the following:

$$\begin{array}{r} \overset{1}{2}71.\overset{1}{2}5 \\ \underline{130.59} \\ 401.84 \end{array}$$

Category 2. Sally reasons as follows:

270 and 130 are 400.
1.25 and .50 are 1.75.
1.75 and .09 are 1.84.
Hence the sum is 401.84.

Category 3. Donna, using a pencil, handles it this way:

$$271.25 > 270$$
$$130.50 > 130$$
$$\text{Hence } 271.25 + 130.50 > 270 + 130 = 400$$

Category 4. Jeremy reasons thus:

271.25 is approximately 270.
130.59 is approximately 130.
270 and 130 are 400.
Answer is approximately 400.

Activities in school mathematics do not fall equally in these four categories. The majority of them fall in category 1; except for learning the number facts and committing them to memory, the majority of mathematical activities involve paper and pencil, and only the exact answer is acceptable. Second most common is category 4. (It is probably for this reason that the terms mental arithmetic and estimation are often used interchangeably.) Activities in category 2 are less common, since mental computation is only rarely used on larger numbers. Similarly, paper and pencil are not needed to find certain estimates; these constitute category 3. Here we are concerned with the two categories in the second column of figure 5.1 (mental arithmetic). The distinction between categories 2 and 4 is not very marked.

The Role of Mental Arithmetic in the School Curriculum

During the past hundred years of mathematics teaching, the role of mental arithmetic has changed considerably. Its emphasis and de-emphasis at various periods has been governed by many factors including the availability of writing materials, the prevailing theories in psychology, and the objectives of teaching arithmetic as perceived during the given period. The last part of the nineteenth century witnessed a strong emphasis on mental computation. Arithmetic was seen as a tool to train and develop the capacities of the mind and the memory. During the first quarter of the present century, the school's emphasis on mental computation decreased but did not completely disappear. The arithmetic books of Thorndike (1924), for example, contain some selected mental exercises at all levels. The stress on speed as well as accuracy in that period forced students to seek shortcuts to do the computations. Hall (1947) noted that during the 1930s less and less work was devoted to mental arithmetic. A review of some recent textbooks shows that this trend, although starting to reverse, still exists.

Different educators have stressed mental arithmetic for different reasons. Spitzer (1967) claims that mental arithmetic improves children's problem-solving ability. Likewise, its implementation forces children to become better acquainted with the structure of the number system. Oral arithmetic helps to guard against the development of meaningless verbal memorization of facts and algorithms. Moreover, on the effective side it motivates students, piques their curiosity, and enhances a cooperative spirit in the classroom.

Most teachers who promote mental arithmetic stress these activities for more utilitarian reasons. Much of the mathematics used by society outside the school tends to be mental, and teachers should make it their responsibility to develop these needed skills. A primary goal of mathematics is to furnish students with tools to cope with the arithmetical demands of the present society. Consider three examples of why mental arithmetic is important in real life.

First, developing skills for efficiency and accuracy in computing with paper and pencil remains a desirable aim in school mathematics. Yet no matter how fast and accurate adults get to be in computing with paper and pencil, everybody gets caught up in situations where these tools are not available: watching a football game, driving a car, attending a cocktail party, and so on. Hand calculators are steadily becoming more accessible, but their use still has some limitation. Until the time comes when these calculators are as accessible as watches, mental arithmetic will remain an important skill in our daily living.

Second, many real-life problems do not need exact answers for their solution. Consider this problem concerning a person shopping at the corner store. Five items are purchased for $1.75, $1.25, $2.15, $1.50, and $2.50. The problem asks whether a ten-dollar bill is sufficient to make the purchase. Even if a calculator is available, estimation is the most efficient method to solve this problem. Finding the exact answer would be a waste of time.

Third, the magnitude of error in some real-life computations could be rather crucial. In these situations it is important to make a quick mental estimate to check whether the results are meaningful and reasonable. For example, in a normal mathematics paper-and-pencil test, Johnny's answer of 322 to the problem 150 + 170 is marked wrong. So is Susie's answer 2120. Both students have followed the addition algorithm with some error. (Johnny reasoned that 0 + 0 is two zeros; Susie forgot to carry the ten 10s to the hundreds position.) In real-life situations, the two mistakes may not be of the same consequence. A person who sends a check for $322 for two items that cost $320 is not "as wrong" as another who sends a check for $2120 for the same two items.

Implications of Teaching Mental Arithmetic in the Middle Grades

The need to develop the skills related to estimation and mental abilities has some very important implications to the practice of teaching middle school mathematics. Probably the most obvious implication is for the teacher to plan appropriate instruction. Just as in any other skill to be developed, students should have ample experiences and practice in situations using these skills. In other words, students need to be taught the different methods that could be used, and they need to be drilled on these methods. It is not sufficient to hope that students would develop those skills when they need them. In every mathematics program there should be a time for the teacher to say, "Put away your pencils and paper and let us solve these problems." NCTM's 1978 Yearbook, *Developing Computational Skills,* gives some specific guidelines for classroom instruction (Trafton 1978, p. 211–13):

- Make procedures and skills in estimation and mental arithmetic an objective for all.
- Recognize that it needs to be taught.
- Build basic number ideas, concepts of operations, and computational skills meaningfully.
- Stress mathematical reasoning throughout computation.
- Make the work an integral part of the instructional program.
- Place estimation and mental arithmetic in the context of application situations.
- Stress oral work.
- Accept individual differences.

However, including mental arithmetic in school mathematics has some further implications to the curriculum in the middle grades. Four implications need to be stressed:

1. *Emphasis on understanding.* Developing the skills related to mental arithmetic and estimation calls for more rather than less mathematical content and more rather than fewer concepts. Through sufficient repetition and drill, students can master the standard paper-and-pencil algorithms without necessarily understanding such concepts as place value, properties of the operations, and so on. However, understanding these concepts is essential to performing arithmetic operations mentally. Take the example 23×70. To do this mentally, a child may reason thus:

$$23 \times 70 = 23(7 \times 10) = (23 \times 7) \times 10$$
$$23 \times 7 = (20 \times 7) + (3 \times 7) = 140 + 21 = 161$$
$$23 \times 70 = 161 \times 10 = 1610$$

It is clear that the following concepts have been used: multiplication by 10, the associative law, place value, and the distributive law.

2. *Emphasis on problem solving.* Developing skills in mental computation calls for the use of algorithms that are often rejected in paper-and-pencil computations. The addition and division problems given at the beginning of this article illustrate the use of the alternative algorithms. Moreover, mental computations use a variety of techniques that are most helpful in certain examples but not in others—for example, to find the square of a number:

$$13^2 = (13 - 3)(13 + 3) + 3^2$$
$$13^2 = 10 \times 16 + 3^2 = 169$$

In general

$$a^2 = (a - b)(a + b) + b^2.$$

In many instances, mental algorithms do not exist or are forgotten. Hence, mental computation also falls under the category of problem solving.

3. *Emphasis on reflection.* The introduction and use of hand calculators in the classroom and everyday life calls for increased mental arithmetic. Provided calculators are available at any time they are needed (as available as watches, which, obviously, they aren't), mental computations are still essential for two reasons. On the one hand, one can argue that if exact answers to computations are needed (figuring taxes, paying bills, etc.), then a calculator is helpful. When only an approximation is sufficient, it can actually be less efficient to use a calculator. Also, even though a calculator may be assumed to be 100 percent error free, the operator, being human, is always subject to error. Suppose that when the number 60.5 is divided by 2.5, an answer of 2.42 was displayed by the calculator. A quick mental division of 60 by 2 and by 3 indicates that the answer is more probably 24.2 than 2.42.

4. *Emphasis on sequencing.* Traditionally mental arithmetic is taught after paper-and-pencil algorithms are mastered. However, as Edith Biggs argues, the ability of children to perform mentally such sums as $68 + 26$ and $38 + 46$ is the crucial test of readiness for practice in written computations with the tens and units (Ewbank 1977; School Council 1966). Practice with mental computation stresses the meaning of computation and makes children appreciate the usefulness of written arithmetic. Too often mathematics teaching is limited to symbols and rules and does not engage the natural thinking patterns of students in approaching numerical situations.

Sample Activities

Activities to develop mental arithmetic are relatively scarce in textbooks and commercially available learning kits. The following short list of activities and algorithms are among those that can be useful to students in performing mental computations and to teachers in providing practice for children in these skills.

1. In a two-digit subtraction problem where the units digit in the subtrahend is larger—for example, $27 - 18$—add units to the subtrahend to make it a multiple of 10 (in this instance, 2), and add the same number to the minuend ($29 - 20 = 9$). Thus, $27 - 18 = 9$.

   - Alternatively, think $27 - 10 = 17$ and $17 - 8 = 9$.

2. In a division problem such as $64 \div 4$, think $60 \div 4 = 15$ and one more 4 is 16.

   - Alternatively, separate the dividend into two easy-to-work-with parts:

     60 − *40* = *24*

     40 is *10* fours.

     24 is *6* fours.

Hence, since $10 + 6 = 16$, $64 \div 4 = 16$.

3. In a multiplication problem such as 3×27, multiply the larger number's units and tens digits separately and add the two products:

$$3 \times 20 = 60$$
$$3 \times 7 = 21$$
$$60 + 21 = 81$$

   - Alternatively, round the larger number (27) to an even number of tens (30) and multiply ($3 \times 30 = 90$). Then multiply the number of units (3) you added to round it ($3 \times 3 = 9$). Subtract the second product from the first to get the answer ($90 - 9 = 81$).

4. In an addition problem such as $275 + 190$, round the number that is close to an even number of hundreds (round 190 to 200 by adding 10). Make the appropriate adjustment to the second number (subtract 10 from 275). The resulting two numbers will be easier to work with: $200 + 265 = 465$.

   - Alternatively, one of the numbers can be separated:

$$275 + (100 + 90) = 375 + 90 = 465.$$

5. To multiply two-digit numbers having the same tens digit and having the sum of their units digits equal to 10—for example, 53×57:

   - The product of the units ($3 \times 7 = 21$) forms the last two digits of the answer.
   - Substitute the next higher number (6) for one of the tens (5), and multiply by the other tens digit ($5 \times 6 = 30$). This product (30) forms the first two digits of the answer.
   - Hence, $53 \times 57 = 3021$.

Conclusions

It is probably a safe generalization to say that school mathematics teachers and textbooks stress paper-and-pencil work to a much larger extent than alternative methods of computation. Moreover, teachers often discourage mental work by insisting that students write down their solution showing (sometimes also justifying) each individual step. Paper and pencil as a computational aid will always be needed. As numbers get larger and more numbers are dealt with, paper and pencil are proper tools to use. Memory and mental computation are like any other skills that are learned: the more they are practiced, the more they develop (Hintzman 1978). Teachers in the middle grades should acquire the courage to venture into new techniques and methods that are outside the usual textbook or syllabus approach. For the reasons presented in this article, variety is not only the spice of life; it is life's vitamins and minerals.

BIBLIOGRAPHY

Additional Ideas for Activities

Judd, Wallace P. *Problem Solving Kit for Use with a Calculator.* Chicago: Science Research Associates, 1977.

Kramer, Klaus. *Mental Computations.* Books A–F, Teachers Guide. Chicago: Science Research Associates, 1965.

Nelson, Glenn, and Larry P. Leutzinger. "Let's Do It without Pencil and Paper." *Arithmetic Teacher* 27 (January 1980): 8–12.

Sample of Research Studies on Mental Arithmetic

Flournoy, Frances. "The Effectiveness of Instruction in Mental Arithmetic." *Elementary School Journal* 55 (November 1954): 148–53.

———. "Providing Mental Arithmetic Experiences." *Arithmetic Teacher* 6 (April 1959): 133–39.

Fuller, Peter W. "Attention and the EEG Alpha Rhythm in Learning Disabled Children." *Journal of Learning Disabilities* 2 (May 1978): 44–53.

Hall, Jack V. "Solving Verbal Arithmetic Problems without Pencil and Paper." *Elementary School Journal* 48 (December 1947): 212–17.

Hitch, Graham J. "The Role of Short-Term Memory in Mental Arithmetic." *Cognitive Psychology* 10 (July 1978): 302–23.

Petty, Olan. "Non-Pencil-and-Paper Solution of Problems: An Experimental Study." *Arithmetic Teacher* 12 (December 1956): 229–35.

Rea, Robert E., and James French. "Payoff in Increased Instructional Time and Enrichment Activities." *Arithmetic Teacher* 19 (December 1972): 663–68.

Sachar, Jane. "An Instrument for Evaluating Mental Arithmetic Skills." *Journal for Research in Mathematics Education* 9 (May 1978): 233–37.

Schall, William E. "Comparing Mental Arithmetic Modes of Presentation in Elementary School Mathematics." *School Science and Mathematics* 73 (May 1973): 359–66.

Sherman, Julia, and Elizabeth Fennema. "Distribution of Spatial Visualization and Mathematics Problem Solving Scores." *Psychology of Women Quarterly* 3 (Winter 1978): 157–67.

REFERENCES

Ewbank, W. A. "Mental Arithmetic: A Neglected Topic?" *Mathematics in School* 6 (November 1977): 28–31.

Hall, Jack V. "Solving Verbal Arithmetic Problems without Pencil and Paper." *Elementary School Journal* 48 (December 1947): 212–17.

Hintzman, David. *Psychology of Learning and Memory.* San Francisco: W. H. Freeman & Co., 1978.

School Council. *Mathematics in Primary School.* London: Her Majesty's Stationery Office, 1966.

Spitzer, H. *The Teaching of Arithmetic.* Boston. Houghton-Mifflin, 1967.

Thorndike, Edward L. *Thorndike's Arithmetic.* Chicago: Rand McNally & Co., 1924.

Trafton, Paul R. "Estimation and Mental Arithmetic: Important Components of Computation." In *Developing Computational Skills,* 1978 Yearbook of the National Council of Teachers of Mathematics, edited by Marilyn N. Suydam. Reston, Va.: The Council, 1978.

6

Interpretations of Rational Number Concepts

Thomas R. Post
Merlyn J. Behr
Richard Lesh

For a variety of reasons rational number concepts are among the most important concepts children will experience during their presecondary years. From a practical perspective, the ability to deal effectively with rational numbers vastly improves one's ability to understand and deal with situations and problems in the real world. From a psychological perspective, an understanding of rational number provides a rich ground from which children can develop and expand the mental structures necessary for continued intellectual development. From a mathematical point of view, rational number understandings are the foundation on which basic algebraic operations will later be based.

Students have consistently experienced significant difficulty dealing with and applying these concepts. Perhaps one reason is that for the most part school programs tend to emphasize procedural skills and computational aspects rather than the development of important foundational understandings. Recent developments in the mathematical, psychological, and instructional realms have revealed important insights into the problems involved in teaching rational number concepts to children.

The cause for concern

The mathematics assessment of the National Assessment of Educational

---

This paper is based in part on research supported by the National Science Foundation under grant number SED 79-20591. Any opinions, findings, and conclusions expressed are those of the authors and do not necessarily reflect the views of the National Science Foundation.

Progress (NAEP) was conducted in 1972–73 and again in 1977–78. It assessed outcomes on mathematics items related to objectives at four cognitive levels: knowledge, skills, understandings, and applications. Commenting on the results of the second assessment, Carpenter and his colleagues (Carpenter et al. 1980, p. 47) made the following generalizations:

> In general, both of the younger age groups [9- and 13-year-olds] *performed at an acceptable level on knowledge and skill exercises. . . . Students appear to be learning many mathematical skills at a rote manipulation level and do not understand the concepts underlying the computation. . . .* In general, *respondents demonstrated a lack of even the most basic problem-solving skills.*

What conditions led to this rather disappointing state of affairs? We suspect the reasons are many, including much premature abstraction of mathematical ideas and a general lack of attention to higher-order thinking skills. A recent examination of three widely used 1978 mathematics text series revealed that by far the greatest emphasis in time spent on rational number concepts (as inferred by the number of pages) is on developing skills with algorithms. This practice continues despite repeated assertions that premature emphasis of algorithmic learning will result in an inability to internalize, operationalize, and apply this concept in an appropriate manner (Carpenter et al. 1978, 1980; Freudenthal 1973; Payne 1976).

Another probable reason relates to the level of abstraction at which much instruction is focused. Children are expected to operate at the abstract/symbolic level too often and too soon. Piaget has suggested that children pass through qualitatively different stages of intellectual development in a predictable order but at varying rates. The stages have been referred to as sensory-motor, preoperational, concrete operational, and formal operational. Children at the age where fractions are normally introduced and developed in the school program will generally be at the concrete operational level. Their ability to synthesize, make deductions, and follow *if/then* arguments very much depends on their personal experience and firsthand interactions with the environment.

Some instructional considerations

Donnelly and Behr (1978) observed that many recent studies have imposed a linear model on Bruner's three modes of representational thought: *first* the enactive phase, *then* the iconic phase, and *finally* the symbolic phase. More realistically, these phases should be incorporated into a nonlinear interactive model. Such a model was suggested by Lesh (1979). This model clarifies certain "translation" processes between modes or phases of representation. The modes of representation identified in the model (see fig. 6.1) are real-world situations, spoken symbols, written symbols, pictures, and concrete (manipulative) models. This model does not imply that one particular mode is always more significant than another. Nor does it

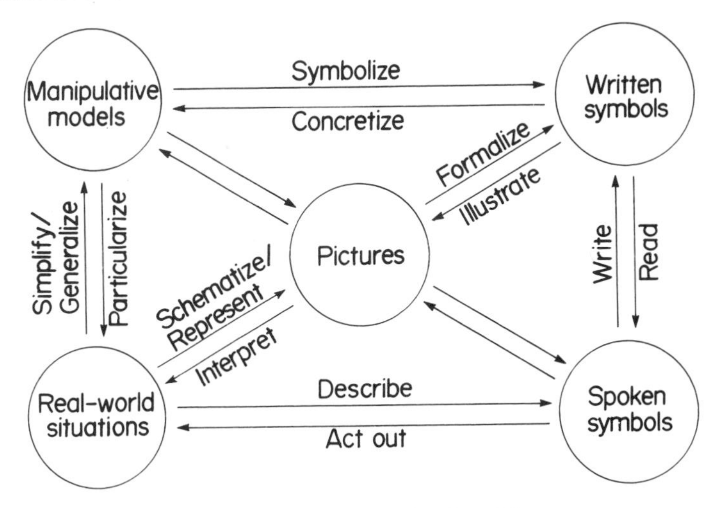

Fig. 6.1. The instructional model

imply that an individual uses all translation processes in a given learning/problem-solving setting.

Often, however, two or more of the identified modes will coexist within a given problem setting or instructional sequence. We hypothesize here that it is the translations *among* and *within* certain of these modes that make ideas meaningful for children. We suggest that this model be imposed on the mathematical interpretations of rational number, which we shall discuss later.

Since understanding a mathematical concept implies the kinds of processes indicated in this model, students should become involved in these translations as they learn about rational number concepts.

Some psychological considerations

Two psychological variables that can be consciously attended to in the instructional process are *abstraction* and *generalization*. The model suggested in this section can be viewed as a bridge between these aspects of conceptual development and the design and implementation of instructional activities.

Dienes (1967) suggests that learning is enhanced when children are exposed to a concept in a variety of physical contexts. That is, their experiences should differ in outward appearance while retaining the same basic conceptual structure. This is known as *perceptual variability*. For example, Cuisenaire rods, geoboards, graph paper, and sets of counters can all be used to depict rational number. By experiencing the concept in a number of physical contexts, the child will be more likely to *abstract* the similarities and discard the irrelevant discrepancies.

Dienes also asserts that *generalization* is enhanced when a concept is viewed from a number of different conceptual perspectives. (This is known as *mathematical variability.*) These two ideas can be incorporated in a two-dimensional matrix (fig. 6.2). Each cell in this matrix implies both a type of physical or symbolic activity *and* a mathematical perspective from which to view the concept. Instructional activities need to be selected so as to give students exposure across both rows and columns. The ultimate value in such a model is that psychological factors (in this case abstraction and generalization) are consciously attended to when constructing instructional activities for children.

MATHEMATICAL VARIABILITY
Kieren's Subconstructs
(Promotes Generalization)

PERCEPTUAL VARIABILITY (Physical Embodiments) (Promotes Abstraction)

	Part/whole relationships	Decimal	Quotient	Operator	Ratio
Counters, egg cartons, and other [set-subset] interpretations					
Rods, number lines, and other *length* interpretations					
Titles, graph paper, geoboards, and other *area* interpretations					
Symbolic algorithm(s)					

Fig. 6.2. Operational definition of the concept of rational number

Some mathematical considerations

The rational number subconstructs that we shall discuss later are not new to school mathematics programs. Kieren (1976) has provided a detailed conceptual analysis of rational number, highlighting hierarchies of subskills within various interpretations of rational number.

The part-whole subconstruct. The part-whole interpretation of rational number is represented in both continuous (length, area, and volume) and discrete (counting) situations. This subconstruct depends directly on the ability to partition either a continuous quantity or a set of discrete objects into equal-sized subparts or sets.

Part-whole applies only when two sets, A and B, are compared and set A is a subset of set B. In addition, the following criteria are satisfied:

1. Set A has been divided into equivalent parts or subsets (in unit fractions this is a single part).
2. Set B has been divided into equivalent parts of subsets.

3. Each individual part or subset of A is equivalent to each individual part or subset of B.

Thus, one can interpret the shaded area of as two-thirds of the whole because the conditions above have been satisfied. That is, the shaded area (set A, part of the whole) consists of two parts that are equivalent to each other. Set B (the whole) has been divided into three equivalent parts. Note also that each part of A is also equivalent to each part of B. In this example, the quantity is *area,* but it could just as easily be volume, length, or number.

It does not make sense to speak of the shaded portion of this triangle as one-third of the area of the triangle, since the parts into which the whole has been divided are not equivalent. This invalidates conditions 2 and 3 above.

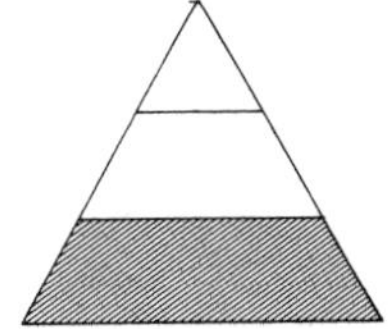

The part-whole subconstruct is an important foundation for other rational number subconstructs, since it is an important beginning point and foundation for children's complete understanding. It is especially useful in developing the naming function of rational number, and it is used to exemplify the relationship between unit and nonunit fractions. For example, looking at the rational number 3/4 (a nonunit fraction) from a part-whole perspective, one easily sees that 3/4 is equal to 1/4 and 1/4 and 1/4.

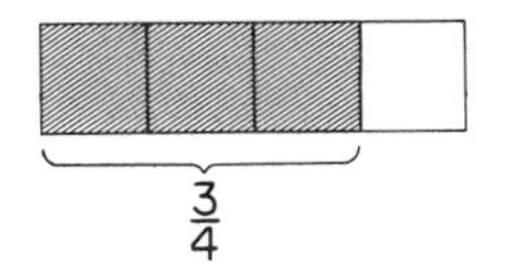

From such statements as 2/4 and 1/4 = $\frac{\square}{\square}$, associated with demonstrations or pictures, children easily generalize to symbolic statements such as 5/13 and 2/13 = $\frac{\square}{\square}$, or 13/72 and 24/72 = $\frac{\square}{\square}$, and finally to $\frac{a}{b} + \frac{c}{b} = \frac{a + c}{b}$.

The association between unit and nonunit fractions also seems to make the transition particularly easy from fractions of a size less than or equal to 1 to those of a size greater than 1. For example, 5/4 becomes 1/4 + 1/4 + 1/4 + 1/4 + 1/4, or 1 whole and 1/4.

The part-whole subconstruct seems to be crucial in providing "preconcept" activity for equivalence and order relations and for operations on rational numbers. The demonstration of the equivalence (or nonequivalence) of fractions based on manipulative materials requires the ability to "repartition" a continuous object or a set of discrete objects.

The multiplication of rational numbers can be introduced as an extension of the multiplication of whole numbers—as the problem of finding a part *of* a part, or a fraction *of* a fraction.

Figure 6.3 shows how paper folding and chips can be used to exemplify $3/4 \times 2/3$.

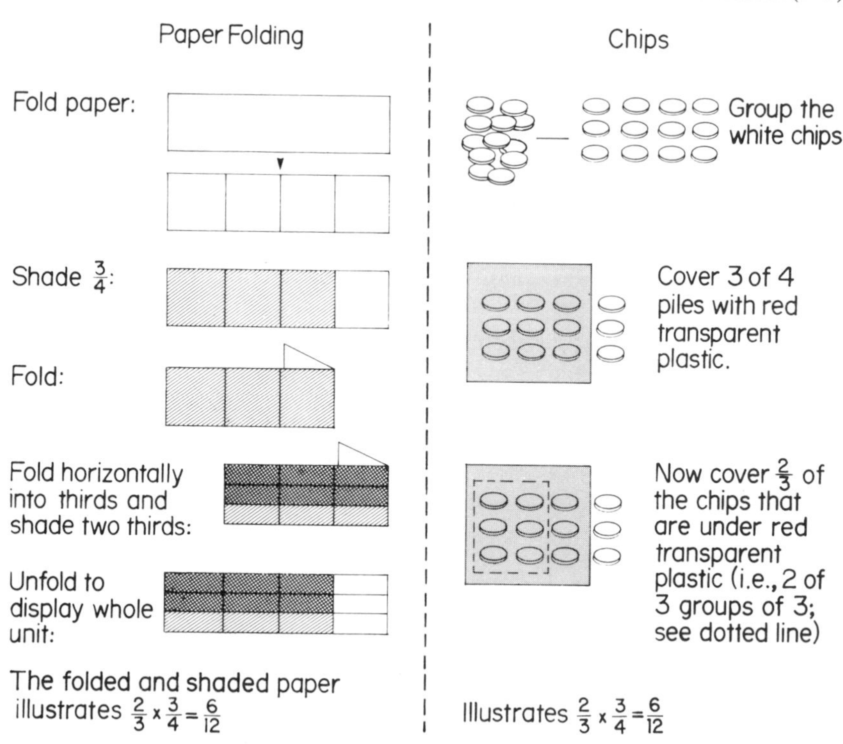

Fig. 6.3. This is an example of perceptual variability. The concept is illustrated here in both a continuous (area) and a discrete (chips) embodiment.

Rational number as decimal fraction. A rational number is any number that can be expressed as a terminating or repeating decimal. The decimal fraction interpretation of rational number is closely related to the part-whole subconstruct discussed earlier. Length (number line), area, and volume, which are referred to as measure systems (Osborne 1976), are useful contexts within which to embed decimal ideas. The symbol system for fractions as decimals is a logical extension of the base-ten numeration system for whole numbers.

A manipulative aid that uses the part-whole subconstruct as a basis for decimals and is similar to paper folding can be made by using a piece of paper one decimeter square as a unit; this unit is partitioned with lines into 10 and 100 equal-sized parts. Figure 6.4 illustrates the aid and shows a representation of 1.36 as 1 + 3/10 + 6/100, or 1.00 + 0.3 + 0.06.

This embodiment is also useful for helping children learn to compare rational number with decimal notation. Figure 6.5 illustrates a comparison of 0.2 and 0.12. From the diagram a child is able to see that more is shaded to display 0.2 than to display 0.12; they can thus conclude that 0.2 > 0.12.

To relate the decimal interpretation of rational number to the number

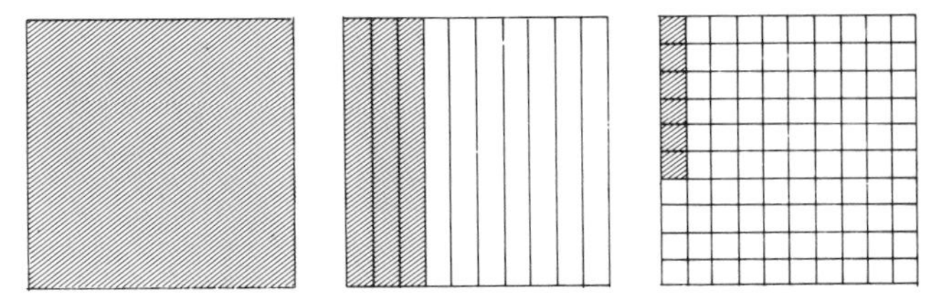

Fig. 6.4. A paper-folding representation of 1.36 based on the part-whole subconstruct of rational number

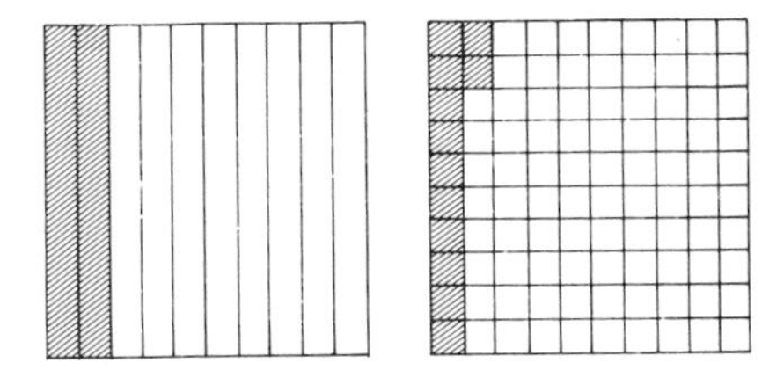

Fig. 6.5. A comparison of 0.2 and 0.12

line, a meterstick is a useful model for both comparing and adding decimals (fig. 6.6).

It is crucial to an adequate understanding of fractions as decimals that children internalize the equivalent meaning of such symbols as 0.36 and 36/100 and 3/4 and 75/100. The importance of this understanding is apparent when children face the task of comparing fractions in terms of relations (less than, equal to, and greater than) and when they must perform opera-

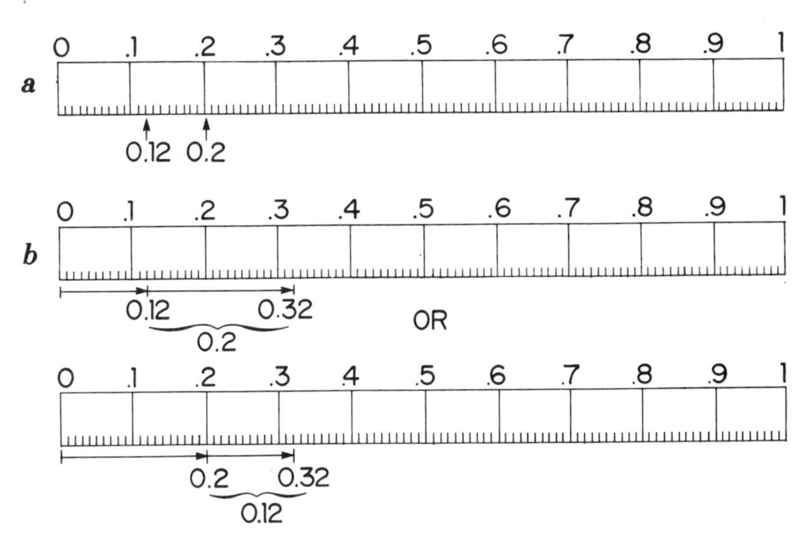

Fig. 6.6. A meterstick is a useful model both for comparing decimals (*a*) and for adding them (*b*).

tions with decimals. Folding a one-square-meter piece of paper will help children visualize the comparative size of products. Figure 6.7 shows how such a unit can help children internalize the comparative sizes of 0.2×0.4 and 0.02×0.4.

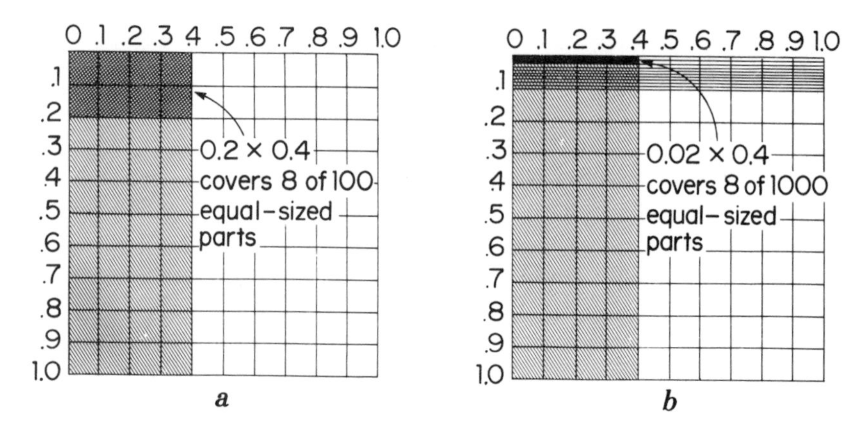

Fig. 6.7. A folded square of paper can help children visualize 0.2×0.4 and 0.02×0.4 and the comparative "sizes" of the two products.

The importance of being able to estimate and to calculate with these estimates in order to approximate the product becomes evident in a problem such as 3.24×9.92: 3.24 is about 3 and 9.92 is about 10; so the product should be about 30.

Rational numbers as indicated divisions and elements of a quotient field. Rational numbers as quotients can be considered at several levels of sophistication. On the one hand, 8/4 or 2/3 interpreted as an indicated division results in establishing the equivalence of 8/4 and 2, or 2/3 and $.\overline{6}$.

On the other hand, rational numbers can be considered as elements of a quotient field and as such can be used to define equivalence, addition, multiplication, and other properties from a purely deductive perspective. The major cognitive structure involved in, and actually generated by, the quotient interpretation is that of partitioning. This notion is also crucial to several of the other subconstructs, but in the case of quotient it is accomplished in a two-stage process, since parts must be recombined following the partitioning process. Consider an example:

> If *a* pizzas are shared equally among *b* persons, how much pizza will each receive?

When *b* does not divide *a,* we have a classic rational-number-as-quotient situation with a nonintegral solution. Thus, the problem of dividing three pizzas equally among four persons can be solved by cutting (partitioning) each of the three pizzas into four equivalent parts

and then distributing one part from each pizza to each individual:

Thus, each person will receive 1/4 + 1/4 + 1/4, or 3/4.

Other situations in which young children may find it useful to think about *a*/*b* as an indicated operation is when an electronic calculator is used to add fractions. For example, if a calculator is used to find 2/5 + 3/8 = □, the problem can be solved using the following steps:

a. Divide 2 by 5 and get 0.4.

b. Divide 3 by 8 and get 0.375.

c. Add 0.4 + 0.375 and get 0.775.

In most problems that youngsters encounter in everyday situations, the answer 0.775 is as useful (or more so) than the answer 31/40, and so there is no need to convert from decimal notation back to fractional notation.

Rational number as operator. This subconstruct of rational number imposes on a rational number p/q an algebraic interpretation; p/q is thought of as a function that transforms a geometric figure to a similar geometric figure p/q times as big, or as a function that transforms a set into another set with p/q times as many elements. When operating on a continuous object (length), we think of p/q as a stretcher-shrinker combination. Any line segment of length 1 operated on by p/q is stretched to p times its length and then shrunk by a factor of q. For example, the function 3/2 would initially triple the length of the unit segment and then halve that length. Thus, the unit length would, under this transformation, become 3/2. Likewise, a length of 4 would be transformed by the operator 3/2 into 12/2 [$(4 \times 3) \div 2$].

In a similar manner, a multiplier-divider interpretation is given to p/q when it operates on a discrete set. The rational number p/q transforms a set with n elements to a set with np elements, and then this number is reduced to $np \div q$.

Figure 6.8a displays a "two-thirds machine" that has accomplished a 2-for-3 transformation. Figure 6.8b suggests that the two-thirds machine can also accomplish a 4-for-6 transformation. That is, two groups of three results in two groups of two. This can be relabeled as 4/6. The two diagrams together suggest that four-sixths machines and two-thirds machines produce equivalent results. It is then concluded that $4/6 = 2/3$.

An interesting situation arises when the machine is directed to operate on 1. The 2-for-3 machine will decompose the unit into three equal parts and emit two of them, that is, $1/3 + 1/3 = 2/3$. Thus, when given one unit as input, the 2-for-3 machine emits 2/3 of that unit. Other fraction machines function similarly.

The function machine will generate equivalent but perceptually different results if it is directed to operate on the whole set or on individual pieces. For example, if we put 3 in a 2-for-3 (two-thirds) machine, then we would get—

a) 2 if the machine works on groups of three (fig. 6.8a);

b) $6 \times 1/3$, that is

$$\underset{\text{1st piece}}{(1/3 + 1/3)} + \underset{\text{2d piece}}{(1/3 + 1/3)} + \underset{\text{3d piece}}{(1/3 + 1/3)}$$

if the machine works on individual pieces. Rearranging, we get $(1/3 + 1/3 + 1/3) + (1/3 + 1/3 + 1/3) = 1 + 1 = 2$.

Rational numbers can be multiplied through a hookup of two function machines. We find $2/3 \times 5/6$ by applying a 2-for-3 machine to the output of a 5-for-6 machine and then finding the single machine that would accomplish the resulting input-output transformation. This is illustrated in figure 6.9. Children can see from the model that the hook-up accomplishes a 10-for-18 transformation, which can also be done with a single 10/18 machine. This suggests that $2/3 \times 5/6$ equals 10/18. In this case, 18 was chosen as the initial

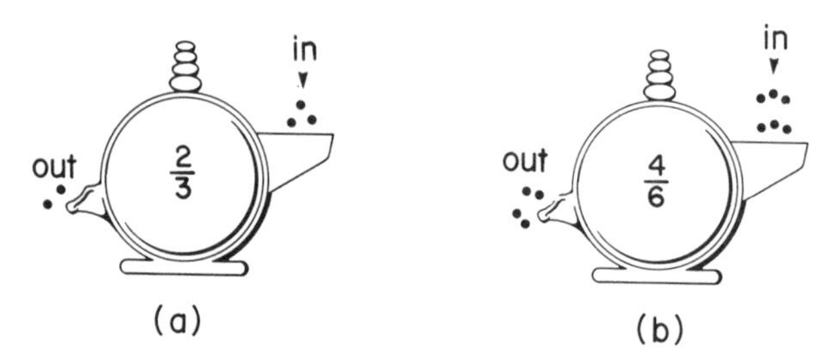

Fig. 6.8. Displays to show how the function-machine model (1) represents the rational number 2/3 and (2) suggests that 2/3 and 4/6 are equivalent fractions

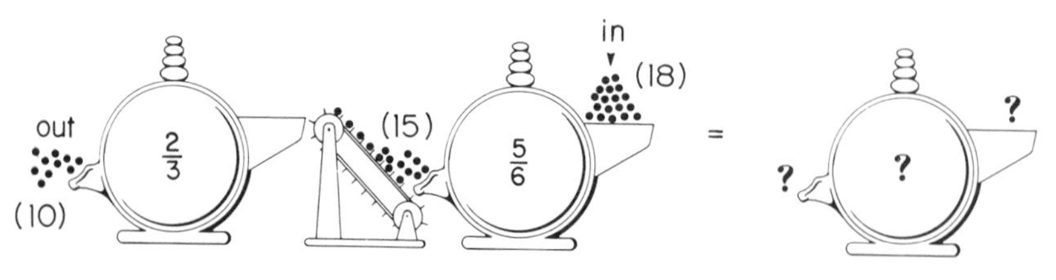

Fig. 6.9. Display to illustrate how the function-machine model is used to embody multiplication of rational numbers

input so that the 2-for-3 machine would have a whole number output. Procedures for finding such "optional" input numbers are analogous to finding common denominators.

Rational number as ratio. Ratio is the special case of a fraction where the equivalent parts of set A are the subsets consisting of the individual elements. Likewise, the equivalent parts of set B are the subsets consisting of its individual elements.

For example, when comparing the set of boys (set A) to the set of girls (set B) in figure 6.10, a comparative index (ratio) of 10 to 20 is generated. Note that 10 to 20 (sometimes written as *10/20,* although *10 to 20* is generally preferred) describes the overall comparative relationship between the numerosity of the two sets.

Fig. 6.10

Equivalent ratios are generated by redefining the individual elements, or comparative unit, and comparing the two sets from a slightly different perspective (fig. 6.11):

Fig. 6.11

Thus, 10 to 20, 2 to 4, and 1 to 2 are equivalent ratios, since each describes a way in which these two sets can be compared. Note that each comparison exhausts all the elements in each set and requires—

a. that set A be partitioned into a number of parts, all of them equivalent;

b. that set B be partitioned into a number of parts, all of them equivalent;

c. that the number of subsets in set A and set B be the same.

A much more common question about ratios is whether one is less than, equal to, or greater than another. When two ratios are equal, they are said to be in proportion to one another. A proportion is simply a statement equating two ratios.

Problems can often be restated in terms of one ratio and half of a second (and equivalent) ratio. The solution amounts to calculating the missing term.

Problem: If candy costs 40 cents for 3 pieces, how much will 5 pieces cost?

This problem can be restated thus:

$$\frac{40}{3} = \frac{N}{5}$$

Ratios can also be considered as ordered pairs of numbers and depicted graphically as points on the Cartesian grid. Points representing equivalent ratios will always lie on a straight line that passes through the origin (0,0). The reciprocal of the ratio represented is, in fact, the slope of the line. Such graphs can be used to generate additional equivalent ratios and to identify the missing element in a proportion. This can be done because the line completely defines the relationship for all ratios equivalent to the given ratio.

Consider again the problem about the candy:

$$\frac{40}{3} = \frac{N}{5}$$

To find N in this proportion, first plot the known relation (3,40), connect with point (0,0), and extend (see fig. 6.12). Now locate 5 (the known part of the second ratio), proceed vertically until the line defining the relationship between cost and pieces is reached, then read the cost directly opposite on the vertical axis (the unknown part of the second ratio).

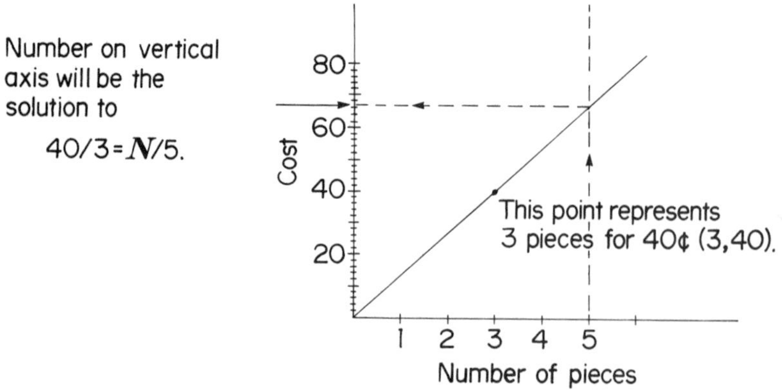

Fig. 6.12

Obviously the equation $40/3 = N/5$ can also be solved symbolically:

$$3N = 5.40$$
$$N = 200/3$$
$$N = 66.\overline{6}$$

The graph, however, is a useful device to illustrate a wide variety of problem settings and to depict problem conditions physically. Incidentally, such experiences with graphing during the middle school years leads nicely into the graphing of more abstract relations, such as $y = 2x + 3$, at a later time.

A final word about learning and manipulative materials

Different materials may be useful at different points in the development of rational number concepts. The goal is to identify manipulative activities using concrete materials whose structure fits the structure of the particular rational number subconcept being taught and ultimately perhaps to fit also the learning styles of individual students.

The magnitude of what is known about how children learn the concept of rational number has increased considerably during the past decade. As new knowledge is gained, additional questions invariably arise. There is still a great deal to learn.

Although rational number has been used to illustrate these points, the basic ideas are believed to apply to the learning of mathematics in general.

BIBLIOGRAPHY

Begle, Edward G. *The Mathematics of the Elementary School*, chaps. 12 and 18. New York: McGraw-Hill Book Co., 1975.

Behr, Merlyn, Thomas Post, and Diane Briers. "Theoretical Foundations for Research on Rational Numbers." In *Proceedings of the Fourth International Conference for the Psychology of Mathematics Education*, edited by Robert Karplus. Duplicated. Berkeley: University of California, Lawrence Hall of Science, 1980.

Bell, Max S., K. Fuson, and Richard Lesh. *Algebraic and Arithmetic Structures: A Concrete Approach for Elementary School Teachers.* New York: Free Press, 1976.

Carpenter, Thomas, Terrence G. Coburn, Robert E. Reys, and James W. Wilson. *Results from the First Mathematics Assessment of the National Assessment of Educational Progress.* Reston, Va.: National Council of Teachers of Mathematics, 1978.

Carpenter, Thomas P., Henry Kepner, Mary Kay Corbitt, Mary Montgomery Lindquist, and Robert E. Reys. "Results and Implications of the Second NAEP Mathematics Assessments: Elementary School." *Arithmetic Teacher* 27 (April 1980): 10–12, 44–47.

Coburn, Terrence G. "The Effect of a Ratio Approach and a Region Approach on Equivalent Fractions and Addition/Subtraction for Pupils in Grade Four." (Doctoral dissertation, University of Michigan, 1973.) *Dissertation Abstracts International* 34 (1974): 4688A–4689A.

Dienes, Z. P. *Building Up Mathematics.* Rev. ed. London: Hutchinson Educational, 1967.

———. "An Example of the Passage from the Concrete to the Manipulation of Formal Systems." *Educational Studies in Mathematics* 3 (1971): 337–52.

Dienes, Z. P., and E. W. Golding. *Approach to Modern Mathematics.* New York: Herder & Herder, 1971.

Donnelly, Martin, and Merlyn Behr. "Review of Research on Manipulative Aids." Duplicated. De Kalb, Ill.: Northern Illinois University, 1978.

Ellerbruch, Larry W., and Joseph N. Payne. "A Teaching Sequence from Initial Fraction Concepts through the Addition of Unlike Fractions." In *Developing Computational Skills*, 1978 Yearbook of the National Council of Teachers of Mathematics, edited by Marilyn N. Suydam. Reston, Va.: The Council, 1978.

Freudenthal, Hans. *Mathematics as an Educational Task*. Dordrecht, Holland: D. Reidel Publishing Co., 1973.

Hartung, M. L. "Fractions and Related Symbolism in Elementary-School Instruction." *Elementary School Journal* 58 (April 1958): 377–84.

Karplus, Elizabeth F., Robert Karplus, and Warren Wollman. "Intellectual Development beyond Elementary School IV: Ratio, the Influence of Cognitive Style." *School Science and Mathematics* 74 (October 1974): 476–82.

Karush, William. *The Crescent Dictionary of Mathematics*. New York: Macmillan Co., 1962.

Kieren, Thomas E. "On the Mathematical, Cognitive, and Instructional Foundations of Rational Numbers." In *Number and Measurement*, edited by Richard A. Lesh and David A. Bradbard. Columbus, Ohio: ERIC/SMEAC, 1976.

Lesh, Richard A. "Mathematical Learning Disabilities: Considerations for Identification, Diagnosis and Remediation." In *Applied Mathematical Problem Solving*, edited by Richard Lesh, pp. 111–80. Columbus, Ohio: ERIC/SMEAC, 1979.

———. "Some Trends in Research and the Acquisition and Use of Space and Geometry Concepts." In *Papers from the Second International Conference for the Psychology of Mathematics Education*, edited by Heinrich Bauersfeld, pp. 193–213. Bielefeld, W. Germany: Institute for Didactics in Mathematics, 1979.

National Assessment of Educational Progress. *The Mathematics Objectives*. Denver: NAEP, Education Commission of the States, 1970.

Noelting, Gerald. "The Development of Proportional Reasoning and the Ratio Concept (The Orange Juice Experiment)." Duplicated. Quebec: Ecole de Psychologie, Université Laval, 1979.

———. "The Development of Proportional Reasoning in the Child and Adolescent through Combination of Logic and Arithmetic." In *Osnabrucker Schriften zur Mathematik*, Proceedings of the Second International Conference for the Psychology of Mathematics Education, edited by E. Cohors-Fresenborg and I. Wachsmuth, pp. 242–77. Osnabrück, W. Germany: University of Osnabrück, 1978.

Osborne, Alan R. "Mathematical Distinctions in the Teaching of Measure." In *Measurement in School Mathematics*, 1976 Yearbook of the National Council of Teachers of Mathematics, edited by Doyle Nelson, pp. 11–34. Reston, Va: The Council, 1976.

Payne, Joseph N. "Review of Research on Fractions." In *Number and Measurement,* edited by Richard A. Lesh and David A. Bradbard. Columbus, Ohio: ERIC/SMEAC, 1976.

Post, Thomas R. "Fractions: Results and Implications from National Assessment." *Arithmetic Teacher* 28 (May 1981): 26–31.

Post, Thomas R., and Robert E. Reys. "Abstraction, Generalization, and the Design of Mathematical Experiences for Children." In *Models for Mathematics Learning*, edited by Karen Fuson and William Geeslin. Columbus, Ohio: ERIC/SMEAC, 1979.

Reys, Robert E., and Thomas R. Post. *The Mathematics Laboratory: Theory to Practice*. Boston: Prindle, Weber & Schmidt, 1973.

Wagner, Sigrid. *Conservation of Equation and Function and Its Relationship to Formal Operational Thought*. 1976. (ERIC: ED 141 117)

Williams, Elizabeth, and Hilary Shuard. *Elementary Mathematics Today: A Teacher Resource* (Grades 1–8). 2d ed. Reading, Mass.: Addison-Wesley Publishing Co., 1976.

7

Problem Solving for All Students

Randall I. Charles
Robert P. Mason
Catherine A. White

MOST teachers at grades 5 through 9 recognize that developing students' problem-solving abilities is one of the most important, yet most difficult, instructional goals to achieve. Our experience at grades 5 through 9 suggests that problem solving can become a significant part of the instructional program for students of all ability or achievement levels. Our ideas are grouped in two areas: *curriculum* and *teaching method.*

Curriculum

In organizing the curriculum for our work, we followed four guidelines: (1) we used both *complex translation* problems and *process* problems for all students; (2) we used activities for all students that developed their problem-solving skills; (3) we provided daily problem-solving experiences for all students; and (4) we posed problems differently for high and low achievers. Let us examine these guidelines individually.

Curriculum Guideline 1	Use both *complex translation* problems and *process* problems for all students.

There are many types of problems one could use in a problem-solving program. We selected complex translation problems and process problems for our work. Below is an example of each.

Complex translation

Carrie is allowed to watch television for 30 hours each week. If she watches for 15 hours on the weekend, how many hours, on the average, can she watch television each weekday?

Translation problems emphasize translating from a real-world situation to a number sentence. *Simple* translation problems involve but one step to arrive at a solution (such as multiplying two one-digit numbers). *Complex* tranlation problems require two or more steps to arrive at a solution and, as a result, usually demand a careful analysis of the story situation. For the translation problem above, here are the steps most students follow to arrive at a solution:

$$\begin{array}{r} 30 \\ -15 \\ \hline 15 \end{array} \text{ hours} \rightarrow \begin{array}{r} 7 \\ -2 \\ \hline 5 \end{array} \text{ days} \rightarrow 5\overline{)15}\ \ 3 \text{ hours each day}$$

Process

Some children are seated at a large round table. They pass around a box of candy containing 25 pieces. Ted takes the first piece. Each child takes one piece of candy as the box is passed around. Ted also gets the last piece of candy, and he may have more than the first and last pieces. How many children could be seated around the table?

Process problems emphasize "the process of obtaining the solution rather than the solution itself" (LeBlanc, Proudfit, and Putt 1980, p. 105). In particular, process problems emphasize (*a*) understanding the problem, (*b*) developing and carrying out a solution strategy, and (*c*) evaluating the solution. Process problems cannot be solved solely by using an algorithm (at least not one known by the problem solver). Usually, the solution of a process problem requires such strategies and skills as organizing information, guessing and checking, drawing pictures, or looking for patterns. Here, for example, is a student's solution to the candy problem; it involved drawing a picture and looking for a pattern:

(2) T X

(3) T X X

(4) T X X X

(5) NO T X X X X

(6) T X X X X X

(7) NO T X X X X X X

2, 3, 4, 6, 8, 12, or 24

Some teachers believe that process problems are not appropriate for low achievers or are too hard for them. On the contrary, process problems not only are appropriate for low achievers but may indeed be the best type of problem to use. The solutions to many process problems can be obtained by those who, for example, have good reasoning ability or organize their work well. Many excellent process problems can be solved without great computational facility. Thus, many students who are low achievers with regard to computation can find success solving process problems.

Curriculum Guideline 2	Use activities for all students that develop their problem-solving skills.

Many students in grades 5 through 9 have not mastered skills that facilitate successful problem solving. For example, many students have difficulty sorting out relevant and irrelevant information in mathematics problems.

Below are four examples of activities to develop problem-solving skills that are appropriate for students in grades 5 through 9.

- Given a problem, determine if there is enough information, extra information, or just the right amount of information needed to solve it.
- Given a situation without a question, make up a question that can be answered.
- Given a problem and an answer, determine if the answer is reasonable.
- Given a problem, determine the conditions in the problem that affect the solution.

In addition to developing problem-solving skills, another reason for using such activities is to give students problem-solving experiences in which some degree of success is almost certain. Most of these activities can be designed so that all students can be praised for some accomplishment in the activity.

Curriculum Guideline 3	Provide daily problem-solving experiences for all students.

Regardless of the amount of time a teacher has available for mathematics instruction, a commitment must be made establishing problem solving as an integral part of the curriculum. For all students, daily problem-solving experiences are useful in developing positive attitudes and abilities related to problem solving. Table 7.1 shows the schedule we followed for our classes. Work with process problems dominated the program, with at least twenty-five to thirty-five minutes twice each week devoted to them (55-minute class periods). All work on a problem was not necessarily completed in one class period. Sometimes students would work on one problem over a period of several days. It is important to note that these time guidelines were not detrimental to our instruction in other areas of the mathematics program.

TABLE 7.1
Time Guidelines for Problem Solving

Content	Frequency	Length
Activities on problem-solving skills	1 day/week	5–10 min/day
Complex translation problems	2 days/week	5–10 min/day
Process problems	2 days/week	25–35 min/day

Curriculum Guideline 4	Pose problems differently for high and low achievers.

The way a problem is stated influences the ideas one generates for solving that problem. Here is a process problem for which most students do not immediately identify a solution strategy.

> What's the greatest number of pieces a pizza can be cut into using 6 straight cuts?

When the problem is stated another way, the problem statement suggests that a *pattern* may exist:

> What's the greatest number of pieces a pizza can be cut into using 1 straight cut? 2 cuts? 3 cuts? 6 cuts?

For low-achieving students, particularly at the beginning of the school year, problems should be posed in such a way that a solution strategy is suggested in the problem statement. Pictures accompanying problem statements, a diagram, and the beginnings of an organized list or table are other ways solution strategies can be suggested in problem statements.

Teaching Method

A teaching strategy for problem solving should be selected to achieve two goals: (*a*) to develop positive student attitudes toward solving mathematics problems and (*b*) to develop students' abilities to select and employ effective problem-solving strategies, such as looking for a pattern.

Developing positive attitudes

The development of students' abilities to solve mathematical problems is seldom accomplished unless a positive attitude toward problem solving exists. Our experience in teaching problem solving at grades 5 through 9 suggests that one of the most important factors influencing students' attitudes about problem solving is the extent to which students are successful in their problem-solving experiences.

Another factor, and perhaps the most important factor influencing students' attitudes, is the teachers' attitude about problem solving. Regardless of the type of problem, the teacher must be careful that the emphasis of problem-solving instruction is on the *process* of solving problems, not the *product,* that is, getting the correct answer. A process emphasis enables all students to find some success in problem-solving situations. Below are three behaviors the teacher should continually recognize and reinforce in problem-solving situations:

- Willingness to attempt mathematics problems
- Perseverance in solving problems
- Selecting a strategy for solving a problem regardless of the elegance or even the usefulness of the strategy (i.e., the strategy selected may not lead to a correct solution or any solution)

Developing problem-solving abilities

Some claim that doing a lot of problem solving is sufficient to develop students' problem-solving abilities and that the teaching actions used in the classroom do not significantly influence problem-solving performance. Our experience, however, leads us to believe that simply doing a lot of problem solving, regardless of how well the curriculum is designed, is not sufficient. Rather, the behaviors or teaching actions used by the teacher can indeed facilitate the development of problem-solving abilities.

The teaching actions described here were used for both process and complex translation problems. Also, most of our problem-solving work was done in small groups of three or four students. For high-achieving students, considerably more independent problem solving was used, particularly late in the school year.

The teaching actions we used can be grouped in three categories: *before* students start work on a problem, *during* work on a problem, and *after* students complete work on a problem. An integral tool in implementing these teaching actions was the problem-solving bulletin board shown in figure 7.1.

Before. Three teaching actions were used before students started work on a problem:

Teaching Action 1	Read the problem to the class, or have a student read the problem, discussing words or phrases students may not understand.

Teaching action 1 was used with all students for the purposes of illustrating the importance of carefully reading problems and of focusing on the meanings of words that have special interpretations in mathematics.

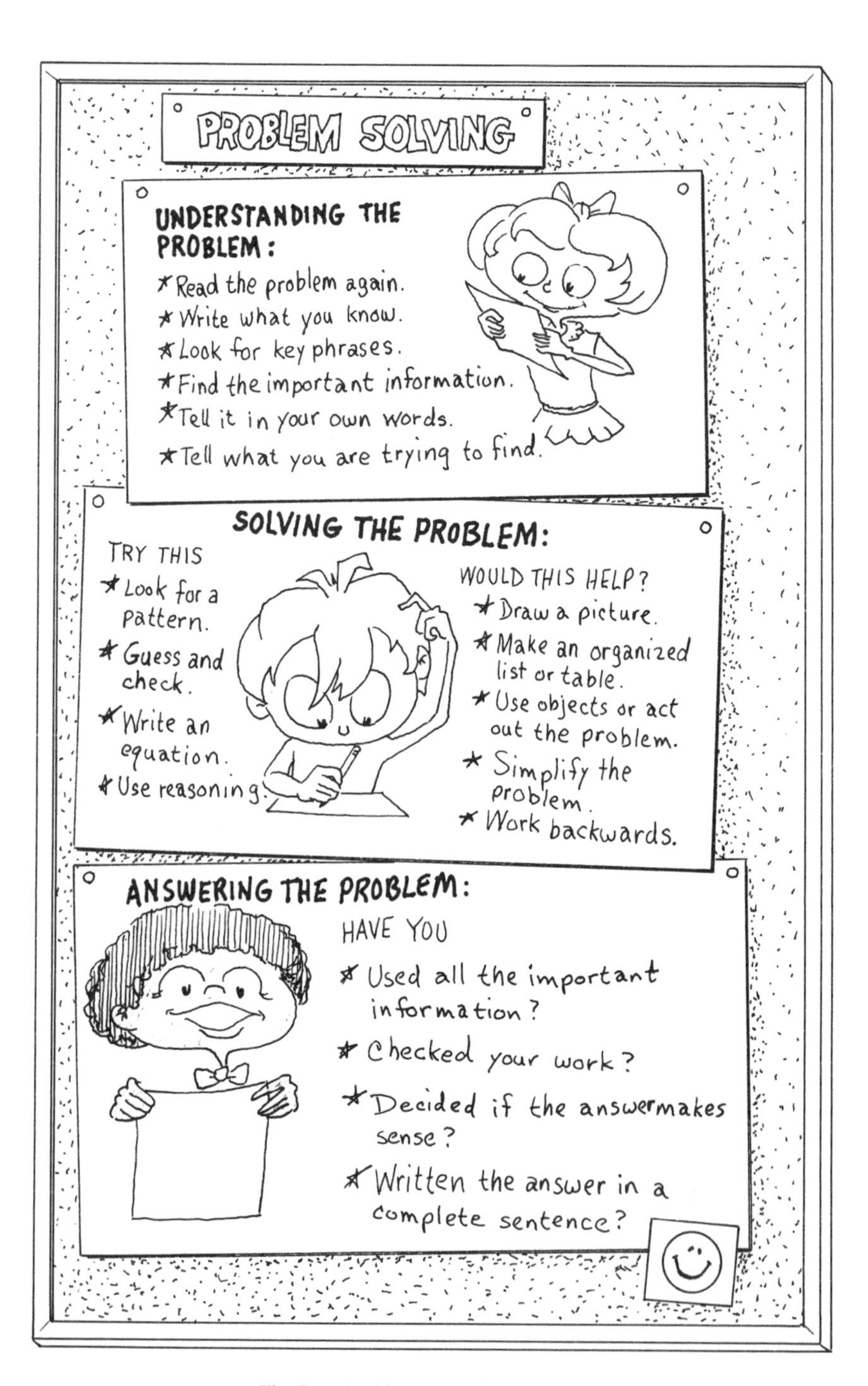

Fig. 7.1. Problem-solving bulletin board

Teaching Action 2	Use a whole-class discussion concerning understanding the problem.

General comments from the "understanding" section of the bulletin board and problem-specific comments should be used to develop an initial understanding of the problem. For the sample problems given earlier, here are some comments the teacher can use related to understanding:

"Tell me this problem in your own words."

"What are they doing with the box of candy?"

"How much television is Carrie allowed to watch each week?"

Teaching Action 3	Use a whole-class discussion with low achievers concerning *possible* solution strategies.

Many low achievers have particular difficulty getting started on problems. To help low achievers, teaching action 3 was used *before* students started work. Following are some comments the teacher can use for the problems given earlier, again using the bulletin board and problem-specific comments.

"Look at the 'solving-the-problem" section on the bulletin board. What strategies might be helpful in solving this problem?"

"Would a picture be helpful?"

"Suppose you had a box of candy. What could you do?"

"Would a calendar be useful?"

During. There are four teaching actions the teacher should use during the time students are working on a problem:

Teaching Action 4	Observe and question students to identify their progress in the problem-solving process.

Teaching action 4 is essential and should be used *throughout* the problem-solving session to facilitate the implementation of the remaining teaching actions.

Teaching Action 5	When students reach an impasse, help them with hints and questions, being more directive for low achievers.

The most critical moment in teaching problem solving is that time when students indicate to the teacher that they're "stuck." That is, they are stymied in their solution efforts and they don't know what to do or try next. To help students past such an impasse, we used *hints and questions* as the basis of our teaching action at this point in the lesson.

Here are examples of some hints and questions for the sample problems.

For high achievers:

"Check the bulletin board to see if you understand the problem."

"Good. You've made a table. Now see if there's a strategy you can use to help you answer the problem."

For low achievers:

"If the box would go all the way around only one time, how many pieces would Ted get?"

"Pretend you have a box of candy. Jim, you are Ted, and Mary is at the table with you. Show me what's happening in this problem."

"When did she watch fifteen hours of television?"

"Write the days of the week, and then write what you know."

It is frequently difficult to identify the reason a student is stumped. Sometimes, for example, students may be attempting to carry out a solution strategy but are having difficulty because they do not fully understand the problem.

Another important factor with regard to teaching action 5 is that the teacher should withhold hints and questions until students have attempted to use the problem-solving bulletin board. One purpose in using the problem-solving bulletin board is to develop students' abilities to help themselves past an impasse rather than relying on the teacher for direction.

Teaching Action 6	Require students who obtain a solution to "answer the problem."

Students who arrive at a problem solution should answer each of the four questions on the problem-solving bulletin board under "answering the problem." All students who obtain a solution should be required to do this.

Teaching Action 7	For students who finish early, give an extension to the problem or have students make up an extension.

This teaching action meets the needs of early finishers by challenging them to generalize their solution strategy or to apply their solution strategy to a similar problem. Here are extensions for the two sample problems:

"Suppose there were seventeen pieces of candy in the box and Ted got the first and last places. How many children could be at the table?"

"If Carrie watched only five hours of television over the weekend, how many hours, on the average, could she watch each weekday?"

After. Near the end of the time students are working on a problem, at least two students should place their solution efforts on the chalkboard. Then a whole-class discussion should be used for the teaching actions in this part of the lesson.

Teaching Action 8	Show and discuss students' solutions using the problem-solving bulletin board as a guideline for discussion.

For high achievers the discussions of solution attempts should be devoted to analyses of solution strategies; discussions concerning "understanding the problem" and "answering the problem" should be initiated only as the need arises. For other students, discussions should be held concerning their work at each stage in solving the problem (understanding, solving, and answering). Questions like these could be asked:

"Which words or phrases were helpful in this problem? Which were confusing? Why?"

"Which strategies, if any—such as drawing a picture—did you use?"

"Can someone give the answer in a different complete sentence?"

Teaching Action 9	If possible, relate the problem to previous problems, and discuss or solve extensions of the problem.

Teaching Action 10	If appropriate, discuss special features of the problem (such as a picture accompanying the problem statement).

Teaching actions 9 and 10 were used with all students to show that strategies used to solve mathematical problems are not problem specific and to illustrate the influence of particular problem characteristics (such as pictures or extra information) on solution attempts.

Conclusion

The terms *high achieving* and *low achieving* have been used throughout this article and have not been defined. Usually, definitions for these terms are developed around some level of performance in previous mathematics classes or on some type of standardized achievement test. Since most mathematics programs and standardized assessment instruments do not at present include experiences similar to those discussed in this article, the terms high achieving and low achieving may not refer to a student's *achievement* with respect to problem solving.

An alternative to viewing students as high achievers versus low achievers is to view them as *experienced* versus *inexperienced* problem solvers. This view suggests that regardless of a student's previous achievement in mathematics, the ideas presented in this article for low achievers may need to be used for most inexperienced problem solvers. If high-achieving students have not had previous experience in a problem-solving program, the guidelines identified for low achievers may need to be used for a period of time. Also, the guidelines for low achievers may occasionally be needed for experienced high-achieving students when extremely challenging problems are used.

The position statement of the National Council of Supervisors of Mathematics (NCSM 1977) and the NCTM's recommendations for school mathematics in the 1980s (NCTM 1980) clearly establish the rationale for emphasizing the development of problem-solving abilities for students of *all* achievement levels. The guidelines presented here represent one way to do this that works!

REFERENCES

LeBlanc, John F., Linda Proudfit, and Ian J. Putt. "Teaching Problem Solving in the Elementary School." In *Problem Solving in School Mathematics,* 1980 Yearbook of the National Council of Teachers of Mathematics, edited by Stephen Krulik. Reston, Va.: The Council, 1980.

National Council of Supervisors of Mathematics. "Position Paper on Basic Skills." *Arithmetic Teacher* 25 (October 1977): 19–22.

National Council of Teachers of Mathematics. *An Agenda for Action: Recommendations for School Mathematics of the 1980s.* Reston, Va.: The Council, 1980.

II. Activities

8

Teaming Up on the Twenties

Frederic R. Burnett

PLACE 150 eighth-grade students in a three-room pod with the partitions removed. Blend in mimeographed materials, rulers, protractors, compasses, and reference materials. Sprinkle lightly with a few teachers. Allow to simmer at room temperature for two days.

What's cooking? Certainly, this recipe would spoil the appetite of just about anyone except the gourmets of team teaching. However, the teachers on my team and I have developed a taste for the team approach and wanted to try a high-impact unit in a combined-discipline setting.

The purpose of the unit was to combine instructional objectives for social studies, language arts, and mathematics into a unit on the 1920s. Emphasizing the economic trends during the twenties, we asked our students to research growth trends, U.S. exports and imports of raw materials and various manufactured and processed goods, participate in a stock market simulation game, and do some creative writing reflecting the tempo of the period.

The specific instructional objectives for each discipline were as follows:

Mathematics

1. Represent data through bar graphs, line-segment graphs, and circle graphs.
2. Calculate percents and percentages.
3. Maintain a ledger sheet for cash transactions.
4. Compute with decimals and fractions.

Social Studies

1. Research events leading to the stock market crash of 1929.
2. Evaluate statistical trends of the economy between 1920 and 1929.
3. Suggest methods for avoiding future stock market crashes.
4. Participate in a stock market simulation game.

Language Arts

1. Write a front page of a newspaper from the 1920s, reflecting newsworthy items of that time.
2. Write a poem with a theme from the 1920s. For example:

> Morning has broken but it's still here.
> A sigh of relief releasing my fear.
> Guess you think I'm in quite a caper.
> No, just that my stock went up in the morning paper.
>
> CAROL WAITS

How the Unit Worked

The structure of the unit was such that our classes were combined for two days, physically as well as academically. We assigned the students a variety of projects along with a continuous stock market game. The game took about fifteen minutes each hour, with each intervening forty-five-minute interval devoted to working on projects. Our basic approach for the projects was to have the students take historical facts from the economy of the twenties and represent them statistically. The data could be represented by bar graphs, line-segment graphs, and circle graphs. This would include graphs of imports and exports, circle graphs showing what portions of the dollar were being spent in which industries, and growth trends of several key industries and stocks (figs. 8.1 and 8.2). After representing the data, the students had to

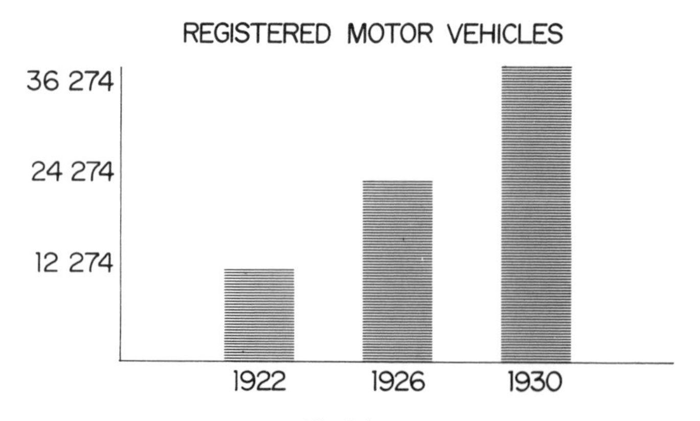

Fig. 8.1

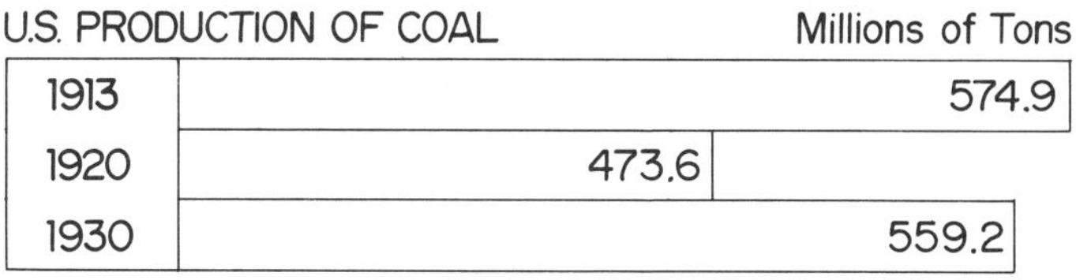

U.S. PRODUCTION OF COAL	Millions of Tons
1913	574.9
1920	473.6
1930	559.2

Fig. 8.2

answer evaluative questions by interpreting the trends in the graph. This served as an excellent tie between social studies and mathematics.

A sample assignment appears in figure 8.3.

Education in 1920 Compared to Today

The educational level achieved by males in 1920 is given in the table. The projected (what we expect) educational level achieved by males in 1980 is also given. Draw a bar graph for each year and compare the two graphs. What conclusions do you reach?

1920 Level Achieved	1920 Percent of Male Population	1980 Level Achieved	1980 Percent of Male Population
No schooling	7	No schooling	1
Elementary	69	Elementary	16
High school	16	High school	56
College	6	College	27

Source: David Popenoe, *Sociology*. 2d ed., (Des Moines: Meredith Corp., 1974), p. 327.

Fig. 8.3

None of the students had much difficulty in drawing bar graphs or line-segment graphs, except in accurately setting up the scales on the axes. But circle graphs were a different story. Accurately setting up sectors of a circle, finding fractions of 360°, using a protractor to determine central angles, and representing various categories of data as percentages of a whole proved sufficiently challenging for our students. The projects assigned by language arts required the students to reflect on the atmosphere of the twenties by writing a poem or making up the front page of a newspaper containing news that would be current to that period.

Stock Market Game

The stock market simulation game was central to the unit. We wanted our students to experience the events leading up to and including the crash of '29. Therefore, using the stock market game seemed natural. We began by making up nine fictitious companies, creating a stock certificate for each (fig. 8.4), and making about one hundred copies of each certificate. Nine students volunteered for the stockbrokers' jobs, which gave them control of the nine

ONE SHARE
PHONY STOCK
DOUBLE TALK
TELEPHONE COMPANY
CERTIFICATE FOR ONE SHARE
Gusher Oil
COMPANY
1 (one) share stock
ONE SHARE
TIN LIZZIE
AUTOMOBILES
ONE SHARE
UNCLEAN
MINING
COMPANY
CHOO-CHOO
RAILROAD
1 Share
STOCK CERTIFICATE
BIG
BROADCASTING
MOUTH
BMB
ONE SHARE
1 SHARE 1
STAR-MAKER-MOVIES
SHADY DEAL
Employment Agency
1 Share
STOCK CERTIFICATE -ONE (1) SHARE
HEAD
IN THE CLOUDS
AIRLINES

Fig. 8.4

respective stocks. Because of the special nature of their assignment, they did not participate in the buying and selling of stock as the other students did. We issued each of the remaining students $100 in play money.

The game was structured so that the nine companies would have price fluctuations over a ten-year period. We began in 1920 with a specific value for each stock. We had a price change for each stock in 1921, again in 1922, and so on each year throughout the game until 1929.

The game started with each of the nine stocks selling for twenty dollars. Thus, the stock market opened in 1920 with about 140 consumers out to purchase stocks from a choice of nine different companies. Because the students were basically unfamiliar with market trends for the various industries during the twenties, their purchases did not necessarily follow any particular rationale. They all sold for the same price at the beginning, and so the students had no real preference for any particular stock. We kept the market open for the 1920 period for about fifteen minutes. At the end of the first trading session, we closed the market, and the students returned to their projects. In about forty minutes, we announced that it was now 1921; after we announced price fluctations and the updated value of the stocks, we opened the market for another round of trading. The students went back to the brokers, bought and sold stocks, and recorded profits and losses on ledger sheets and line-segment graphs. And so the day progressed, with the stock market opening for trading five times during the day, each opening corresponding to a year. The first day took us through 1924, with each year showing price fluctuations coming from historical events that had occurred during that year.

We began the second day by opening the market for trading; it was 1925. By this time the students were forming opinions on the trends and potentials of the stocks and were demonstrating this by being more discriminating in their purchases. Also, word had gone around about the impending crash, and we were getting the first undercurrents of imminent doom. During this second day, we also had the market open five times, and student interest and involvement increased with each successive trading period. For example, in 1926, we had the auto company rise fifteen points, an event foreseen by some of our more insightful students. Announcing the increase for this stock brought a rousing cheer from our group of wise investors. The undercurrent was getting stronger. In the meantime, one of the nine stocks, an employment company, had dropped from an original value of twenty dollars a share in 1920 to five dollars a share in 1927. Since the various research projects had hinted at widespread depression in the 1930s, many of our students felt that it would be profitable to invest now in the employment company. Their reasoning was that the depression would send droves of customers to the employment company, thus sending the value of the stock sky-high.

Well, in 1928 it started happening. The market opened, and the students wanted to cash in their prosperous industrial stocks for profit and invest everything in the employment company. In no time at all, the stocks for the employment company had sold out, and we had quite a few disappointed people in the pod. However, this disappointment was minor compared to that of the students who wanted to sell their stocks.

I must point out here that we started the game by issuing each broker about $1000 for transaction purposes. By this time, however, the stocks had risen in value to a point where the currency in circulation could in no way cover the increased value of the stocks. Therefore, after the first few transactions in 1928 were completed for each broker (that is, after they had redeemed the customers' stocks), there was no more cash! The result was that a majority of the students were left high and dry. With no cash to back up the value of the stocks, their certificates were worthless. The net result was *panic.* Who could have anticipated that the value of the stocks would be inflated and that there would be no real money to back up the certificates? Thus, their disappointment in not being able to convert stocks to cash really simulated (if not imitated) the crash of '29. Students ran around in confused amazement; feeble attempts at wheeling and dealing proved fruitless. (There were even a couple of attempts at passing counterfeit cash, but this was quickly squelched by the other students and the teachers. We had enough problems with worthless stock certificates.)

The ensuing chaos of anticipating the crash prompted us to bring the game to a close, and so we called for everyone's attention while we tallied the final figures of the game. We asked the brokers how much money they had. Most of them had none. With that, we informed our students that the stocks which had no cash to back them up were worthless, and their value for 1929 was zero. A couple of companies still had token reserves of cash, and so they received a token value on the board for 1929. The employment company had risen in value as expected, but most of the students who had that particular stock knew it was the end of the game and had cashed in their certificates as long as the cash held out. Most people showed a profit on their ledger sheet, but very few had the actual cash to show the profit their records said they had made.

One student's ledger sheet, showing a record of her stock transactions and her profits and losses, appears in figure 8.5. Figures 8.6 and 8.7 chart the performances of two of the stocks.

The lessons in the unit were very effective. The specific objectives for social studies, language arts, and mathematics were generally met. But the stock market simulation game was really the main event of the two days' work on the twenties. The causes and events leading up to the crash of '29 were well learned by the students. They had lived through it! In fact, after the panic selling and widespread chaos at the end of the decade, all of us now give a new meaning to the term Roaring Twenties.

Although our game structure was designed for middle school, teachers should have no problem applying the principles to other classroom settings. If the game is to be applied to a self-contained class, simply do what has already been suggested and have the students break down into smaller groups for the various activities. The teachers can even vary the depth of each objective for the different levels within the class. If the classes are departmentalized, the students can work on the respective disciplines during the appropriate class period, with each teacher handling a portion of the stock market simulation game.

Year	Stock	No. of Shares	Purchase Price	Market Value	Profit/Loss
20	Tin Lizzie Auto	1	$20	$20	0
20	Gusher Oil	1	$20	$20	0
20	Shady Employment	1	$10	$10	0
21	Tin Lizzie Auto	1	$20	$30	$10
21	Gusher Oil	1	$20	$30	$10
22	Tin Lizzie Auto	1	$20	$50	$30
22	Gusher Oil	1	$20	$35	$15
23	Gusher Oil	1	$20	$30	$10
23	Tin Lizzie	1	$20	$65	$45
23	Shady Employment	1	$10	$ 5	−$ 5
23	Shady Employment	3	$ 5 each	$ 5	0
23	Star Maker Movies	1	$30	$30	0
24	Star Maker Movies	1	$30	$40	$10
24	Shady	1	$10	$10	0
24	Gusher	1	$20	$20	0
24	Auto	1	$20	$50	$30
24	Shady	3	$ 5 each	$10	$15
25	Shady	3	$ 5 each	$ 5 each	0
25	Shady	1	$10	$ 5	−$ 5
25	Gusher	1	$20	$35	$15
25	Auto	1	$20	$55	$35
25	Star Maker	1	$30	$45	$15
26	Star Maker	1	$30	$45	$15
26	Gusher	1	$20	$45	$25 sell out
26	Shady	1	$10	$ 5	−$ 5
26	Shady	3	$ 5 each	$ 5	0
26	Auto	1	$20	$70	$50 sell out
27	Shady	1	$10	$15	$ 5
27	Shady	3	$ 5 each	$15	$30
27	Star Maker	1	$30	$70	$40
27	Big Mouth Broadcasting	1	$65	$65	0
28	Big Mouth	1	$65	$55	−$10
28	Shady	1	$10	$10	0
28	Shady	3	$ 5 each	$10	$15
28	Star Maker	1	$30	$75	$45
29	Shady	1	$10	$ 5	−$ 5
29	Shady	3	$ 5 each	$ 5	0
29	Star Maker	1	$30	$75	$45 sold out
29	Big Mouth	1	$65	$60	−$ 5 sold out
30	Shady	1	$10	$80	$70
30	Shady	3	$ 5	$70	$195

Fig. 8.5

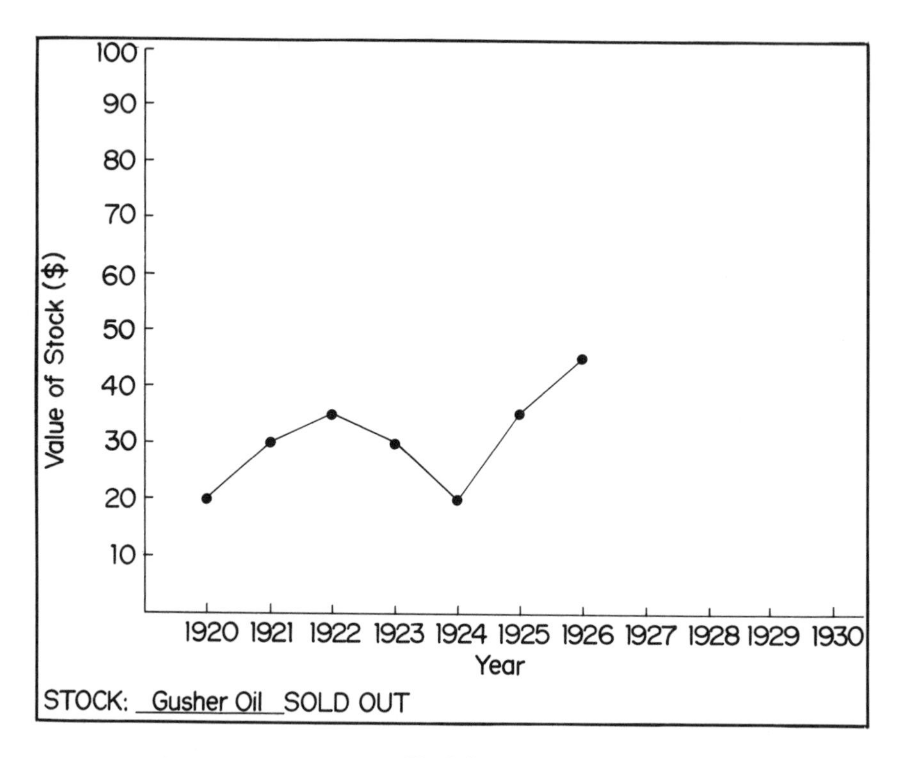
100
90
80
70
60
50
40
30
20
10
Value of Stock ($)
1920 1921 1922 1923 1924 1925 1926 1927 1928 1929 1930
Year
STOCK: Gusher Oil SOLD OUT

Fig. 8.6

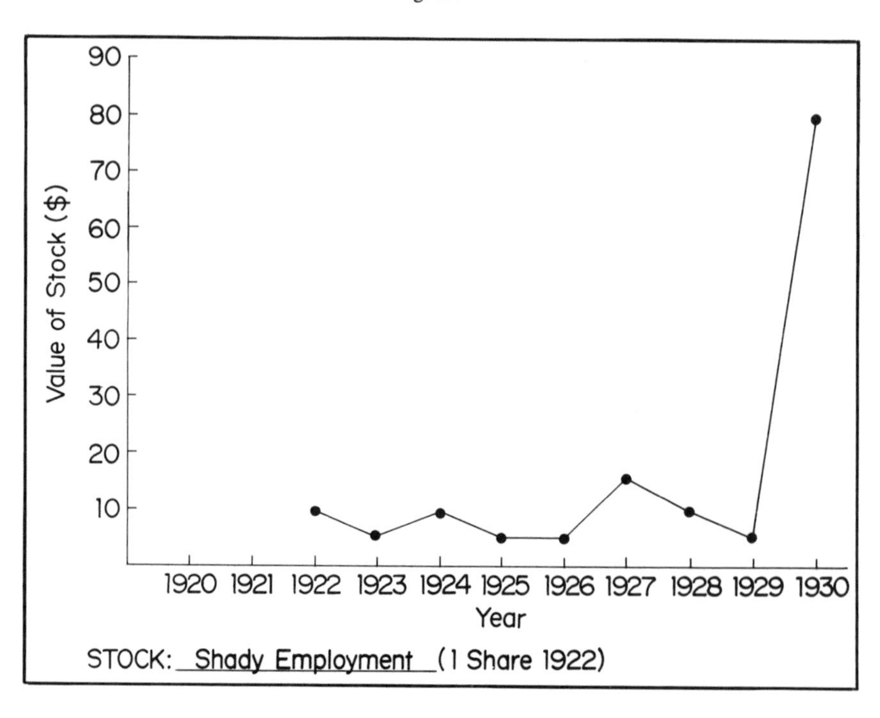
90
80
70
60
50
40
30
20
10
Value of Stock ($)
1920 1921 1922 1923 1924 1925 1926 1927 1928 1929 1930
Year
STOCK: Shady Employment (1 Share 1922)

Fig. 8.7

9

Beyond Four Walls: Mathematics in the Out-of-Doors

Rita J. Casey
with the collaboration of ***Evelyn J. Sowell***

MATHEMATICS in the out-of-doors? What if you don't teach on the edge of a forest and the nearest greenery comes through a crack in the pavement? Even so, almost every mathematics teacher has the opportunity to expand his or her classroom to include something outside the walls of the usual school setting. Many schools have patios, courtyards, playing fields, or recreation areas that can be used for mathematics instruction.

We believe that there are strong motivational advantages to conducting some mathematics activities outside. Outdoors is often the "real world" for students in a way that classrooms can never be. Furthermore, outdoor mathematics offers fun, a departure from routines, and contact with mathematics in a tangible manner.

Although almost any mathematics topic could be taught outdoors, it is important to do more than simply move the books, papers, and pencils out of the classroom. The purpose from an instructional viewpoint is to use the out-of-doors as a suitable setting for active, real-life learning experiences. Often outdoor experiences are ideally suited to strengthen or extend students' understanding of mathematics skills. Certain skills that are particularly appropriate for this kind of setting include measuring, mapping, graphing, and sampling, topics found in several middle school mathematics texts. There are many others as well.

How does it really work? Let's look at a few examples using the four topics just mentioned.

The photographs in this article are by Winston Green and Ben Ferrell, of the University of Texas at Tyler, and Mark Muckelroy, of the Tyler Independent School District.

Measuring

Outdoor measurement activities may help a few students realize that measurement really involves a thing to be measured and something to measure with. These students will begin to understand the meaning of measurement. For others, the outdoor activities offer opportunities to measure larger distances, areas, or objects than they've ever before experienced. A 100-meter race takes on new meaning for those who actually measure off 100 meters.

The outdoors provides situations in which students can understand and use unfamiliar measuring devices. Pushing a measuring wheel is often a completely new experience for students. A pedometer gives learners information on measuring linear distances without applying the measuring tape to the distance covered. How many students get to use a clinometer within a classroom? Yet outside it has all sorts of applications, from measuring the angles that a tetherball makes at various speeds to gauging the angles of shadows.

Being outside can motivate students to use or derive formulas for distance, speed, area, or volume. Staking off several irregular areas and using a measuring wheel, flexible tape, or even a length of cord can help establish meanings for perimeters. If tree stumps or other cylindrical objects are present, students can measure circumferences and diameters. By comparing the ratios of these measures, they can learn the meaning and significance of pi.

Mapping

Usually some facility with measuring devices is required for mapping activities. Students can make maps of certain areas or learn to interpret maps that have been made of an area. Most teachers would agree that the concept of ordered pairs of numbers takes on a lot more excitement for students when they are trying to find the coordinates of a missing object that is to be located using a buried-treasure map. Maps involve ratios in ways that textbook problems do not. Differences in map scales are very dramatic when students actually explore the area represented.

Mapping activities also can involve making or interpreting contour lines, helping students understand the relationship between changes in incline and the distances between contour lines. Orienteering is a sport of increasing popularity that involves the use of contour maps for traveling on foot to specific locations. It's very enjoyable to plan and follow a course on a contour map.

Graphing

Graphing activities are closely related to mapping. The outdoors can be used as a place to collect data for graphing activities. All kinds of graphs are possible, including bar, circle, and line graphs. Students may count the number of times they see certain birds, plants, trees, insects, colors of leaves, and so on, as the basis for bar graphs. Students who really want to help the environmental cause can clear litter from a defined area (exercise caution on the kind of litter to be collected!) and bring it to a central location for classifying before they dispose of it properly. They can use the categories and the percentages of the total number of pieces as the basis for a circle graph. Students will multiply each percentage by 360 degrees to find what portion of the circle should represent each category of litter. Since students are outside, they may engage in some physical fitness routines to count pulse or respiration rates. These data can be used to make line graphs, showing how heart or respiration rates change at fixed intervals during a course of physical activity.

Sampling

Being outside presents many opportunities to teach sampling concepts. First explain to the students how they are to deal with sampling and then let them establish what population they are interested in. Sampling the number of acorns, pine needles, ants, or other small items in a particular measured area is convenient. A wire coat hanger bent into a square gives the students a tool to take samples of the objects found within its area (approximately 400 square centimeters).

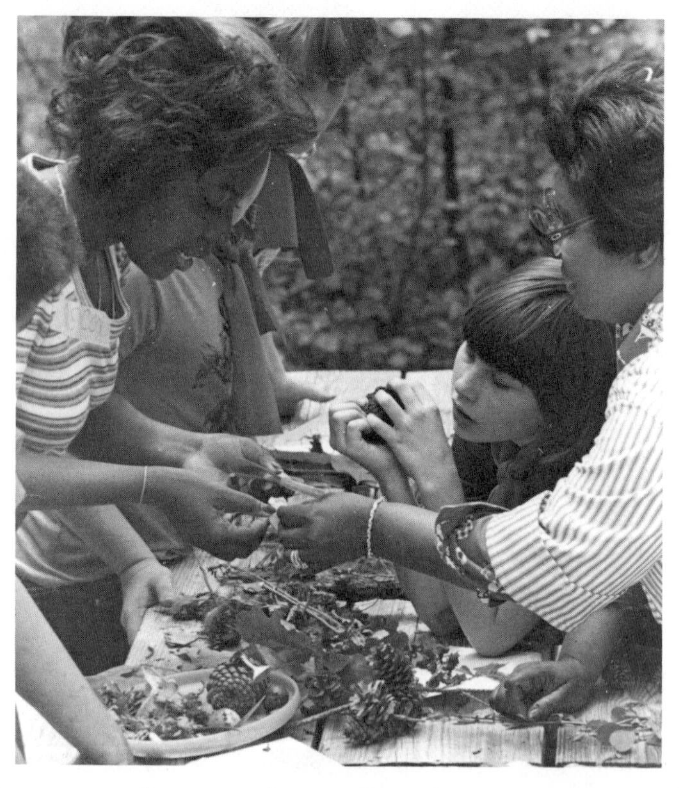

Students may prefer to use a larger unit, such as a square meter, if they're sampling large objects. They may be able to go beyond the usual content and develop the ideas of random or stratified random sampling through outdoor activities (Hays 1981). Students should be encouraged to compare their information in some concrete way with that of other students. This shows them the differences that can occur when samples are taken.

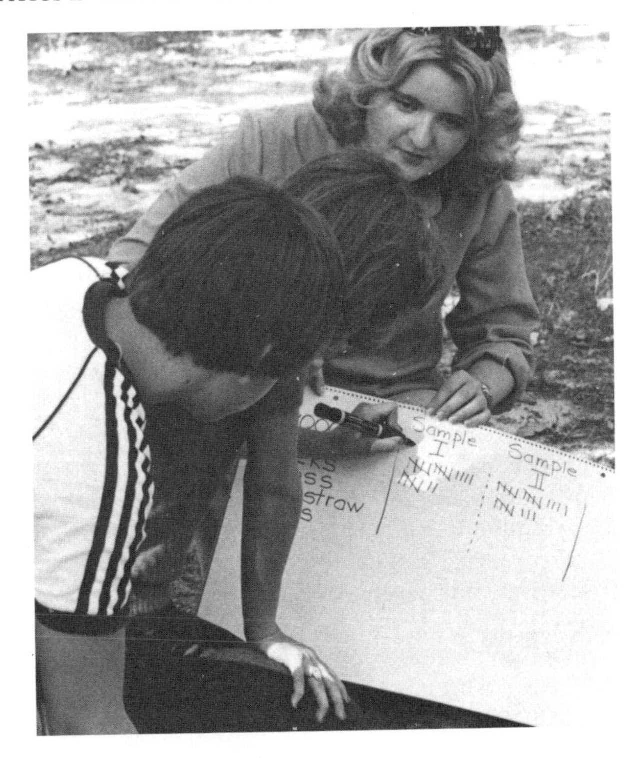

Variations to suit ability levels

The preceding discussion has offered several brief descriptions of activities that can be done outside. These activities can be adapted to students' abilities with only a little planning. For example, if some students are relatively new to measuring, focus their lesson on nonstandard units of measure and gradually build toward the idea of using standard units for communication with other people. Students with little or no mapping skills may first be asked to interpret very simple two-dimensional maps. More advanced students can be asked to construct two- and even three-dimensional maps of given areas. Students with little experience in graphing may need to graph frequencies of discrete objects prior to graphing things of continuous measure. Circle graphs may be approximated, using the students' knowledge of fractional parts of a whole, or constructed with a protractor, depending on the ability of the learners. Sampling activities can range from very simple to very complex. Students may be asked to sample in a relatively small area or over a wide range. The concept of random sampling is one that most students can handle, but stratified random sampling may be reserved for the more advanced students who already have some knowledge of sampling. The ability levels and interests of the students themselves may suggest the kinds of activities they would like to pursue.

Additional considerations

Before the students go outside for a lesson, think through some concerns that might be taken for granted in "familiar" classroom situations. Is learning basically a passive endeavor, with students receiving the teacher's output and reacting to it? Or do pupils interact with things in their learning environment according to their own internal purposes? Mathematics experienced outdoors will be an interactive process. These ideas may not be suited to the personal teaching style of one who is more comfortable with very directive instruction.

Two related concerns are the degree of supervision required and the use of time in the out-of-doors. An activity that requires very close teacher support may be better done indoors. Outside activities need to be well planned and explained at the outset so that students can do their work with a minimum of supervision. Timing must be given very careful attention in the planning and execution of these strategies. It's probably best to do things that can be completed in one outdoor session if a strict period-by-period timetable must be followed. Moreover, since students' work rates may change drastically outdoors, early and late finishers must be taken into account to cut down on managerial difficulties.

Often it is helpful to make outdoor experiences a joint effort by getting another adult to help. This may range from a parent volunteer to a team effort with a fellow faculty member, perhaps a science teacher or even an English teacher. Clearing a larger-than-usual block of time might be arranged with other teachers and administrators. The possibilities are varied.

Perhaps a local high school could provide student helpers from its Future Teachers of America chapter. If a nearby university has a teacher education program, a preservice teacher or two might be able to help. In one instance several classroom teachers and university students took more than a hundred students to a nearby state park for an entire day of intensive outdoor learning. Who says there can't be field trips for mathematics, too? In any event, whether plans are on a large or small scale, an additional person can definitely enhance the quality of an outdoor mathematics experience.

We have found mathematics in the out-of-doors to be a very rewarding teaching experience. Students are eager and ready to go the next time. Although it may take a little more planning than usual, we believe you and your students will have fun and learn a lot, too!

REFERENCE

Hays, W. L. *Statistics for the Social Sciences.* New York: Holt, Rinehart & Winston, 1981.

10

Math Mapping Begins at Home

Bunny Crawford Hatchett

USING a local map to develop map-reading ability for middle school mathematics students is the fundamental approach taken in this article. Even though map study is more in the domain of the geography curriculum, a map unit can provide an interesting change of pace for mathematics students and bring about a noticeable improvement in their map-reading ability.

For a foundation, the students should be introduced to a simple first-quadrant coordinate system and learn to associate number pairs with points on a graph. Two class periods might be spent drawing graphs, learning the vocabulary associated with a coordinate system (*horizontal axis, vertical axis, origin, number pair, coordinates,* etc.), plotting number pairs correctly, and naming coordinates of points on graphs. Introducing graphs early in the year will assure coverage of the topic even for classes that move slowly. Students who subsequently advance to graphing functions will need only a brief review before progressing to a four-quadrant plane. The students should be shown how a graph can represent a city and how to use directions to indicate the positions of the square blocks and parallel streets.

Starting "at Home"

Viewing a city map, the class locates major streets, thoroughfares, state and national highways, and public places such as the courthouse, city hall, schools, parks, shopping malls, fast-food restaurants, and theaters—places that are familiar to all students. Coordinates and street names are used to pinpoint these places, and then students work to see who can name the school in the K–8 block, the park in the H–8 block, and so on (fig. 10.1). A transparency for the overhead projector and an enlarged city map for the

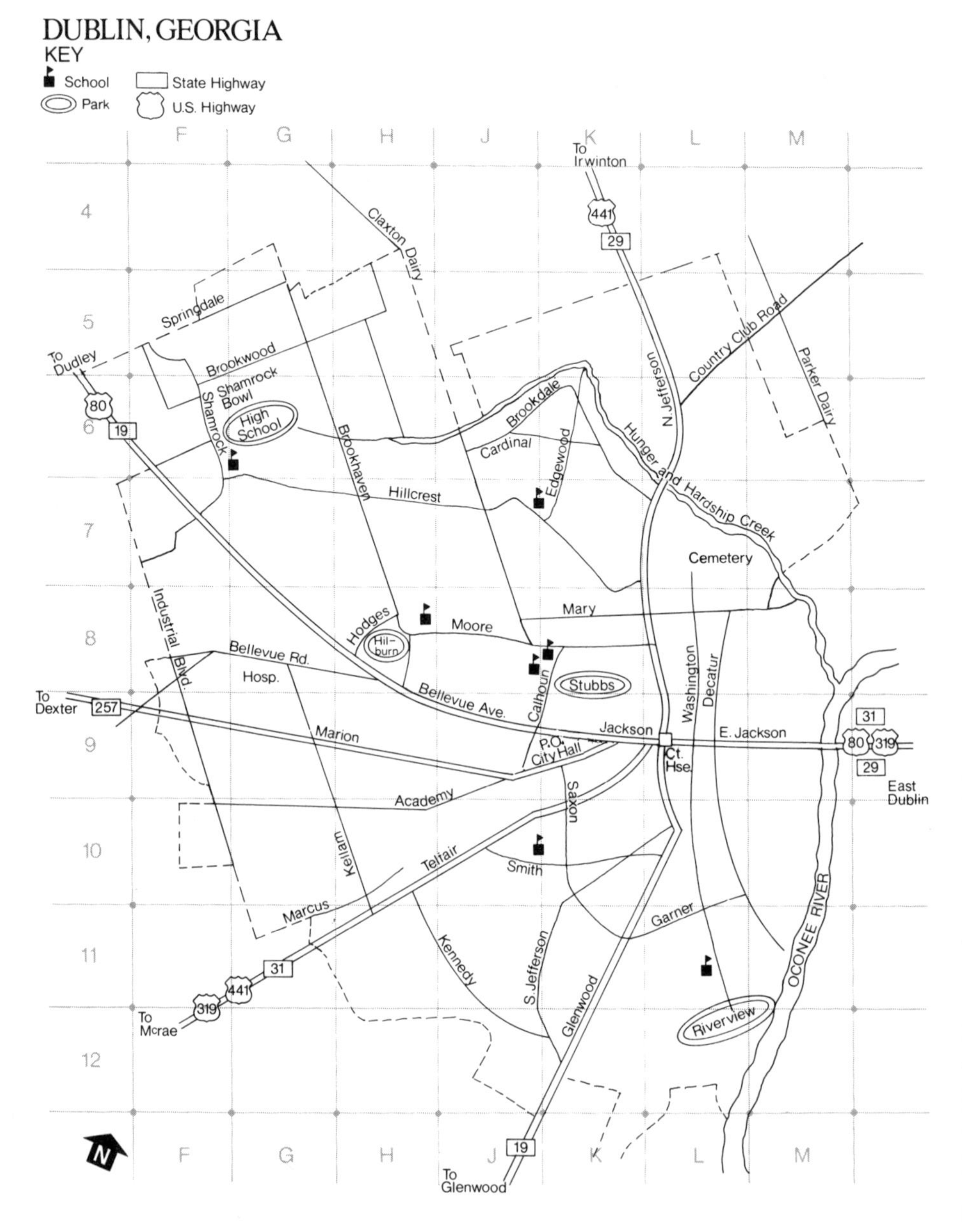

Fig. 10.1

bulletin board (both can be teacher made) can aid classroom discussion. These maps will not show the details of the student maps—only the major thoroughfares, coordinates, city boundaries, and natural features. Highlights of local history and the origin of street and subdivision names may be presented to increase the students' knowledge of their town.

Each student, with some assistance from the teacher or another student, should find her or his own place of residence on the large bulletin-board map and initial the spot. The students can then practice using directions to "travel" from one location to another, asking and answering such questions as "In what general direction would you travel from the city hall to the bowling alley?" Each student should be required to write the directions from school to home using correct street names, directions, and approximate distances, noting landmarks en route. Each description is checked by the teacher; if found to be in error, the paper is returned, pointing out where the directions actually led so they can be rewritten. The importance of giving accurate directions is thereby stressed.

The mileage scale on the map is noted, and students learn to determine approximate distance by comparative measurement, duplicating the scale on the edge of a separate sheet of paper. This method is practical, for often a paper edge is more accessible than a ruler.

Map skills are one of those strategic "life skills" that seem difficult for some students to grasp. Maps often are not relevant to students who rarely travel. They find it difficult, if not impossible, to comprehend such ideas as latitude and longitude and to relate to oceans, deserts, and mountains. Seasoned travelers often have not seen their everyday world on a map and profit from the opportunity to view and work with a map of "home." Most students are able to grasp the idea that a city map is just a small aerial picture of their town, giving both coordinates and directions with increasing ease.

Moving On

As the class progresses to a county map, they might use directions and the mileage scale in planning routes to travel between towns (fig. 10.2). The scale is used to teach ratio solutions for distance problems. Questions such as "How far is it from Dublin to Rentz" or "What church is approximately six miles northeast of Dudley?" should be asked. This allows students an opportunity to use rulers and ratios to measure distances and calculate actual mileage. Geographical features, points of interest, and the interstate highway should be pointed out. Aerial photographs used by agricultural agencies could prove effective here as a visual aid.

Next comes a venture with a state map. The teacher, on introducing this topic, notes the boundaries, the major metropolitan areas, and the interstate highway system and points out the location of the home county and city (fig. 10.3). Satellite maps could enhance the aerial aspect of the map study at this point. After orientation, the study should concentrate on two specific areas: (1) understanding and using the symbols of the map key and (2) finding distances between cities using (*a*) the highway mileage, (*b*) the mileage in the cross-reference of large cities, and (*c*) the airline mileage from the scale of miles and a teacher-furnished metric scale. Finding these distances makes

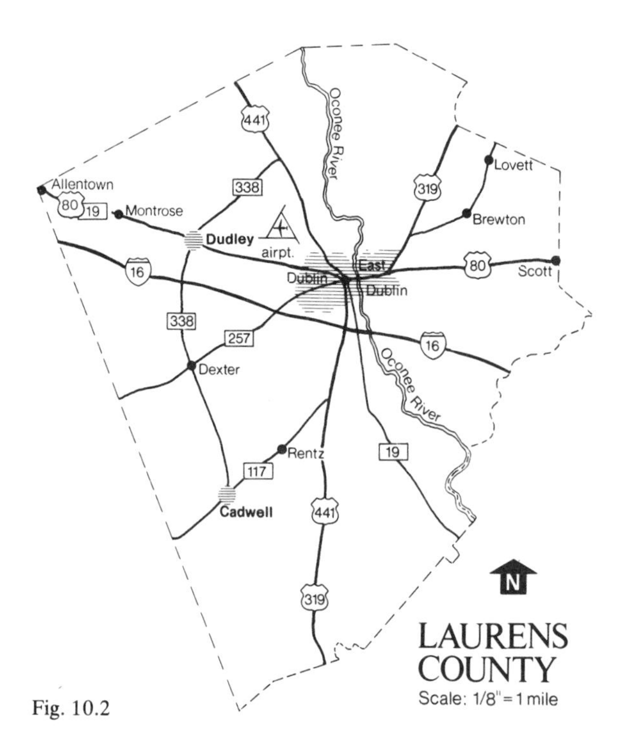

Fig. 10.2

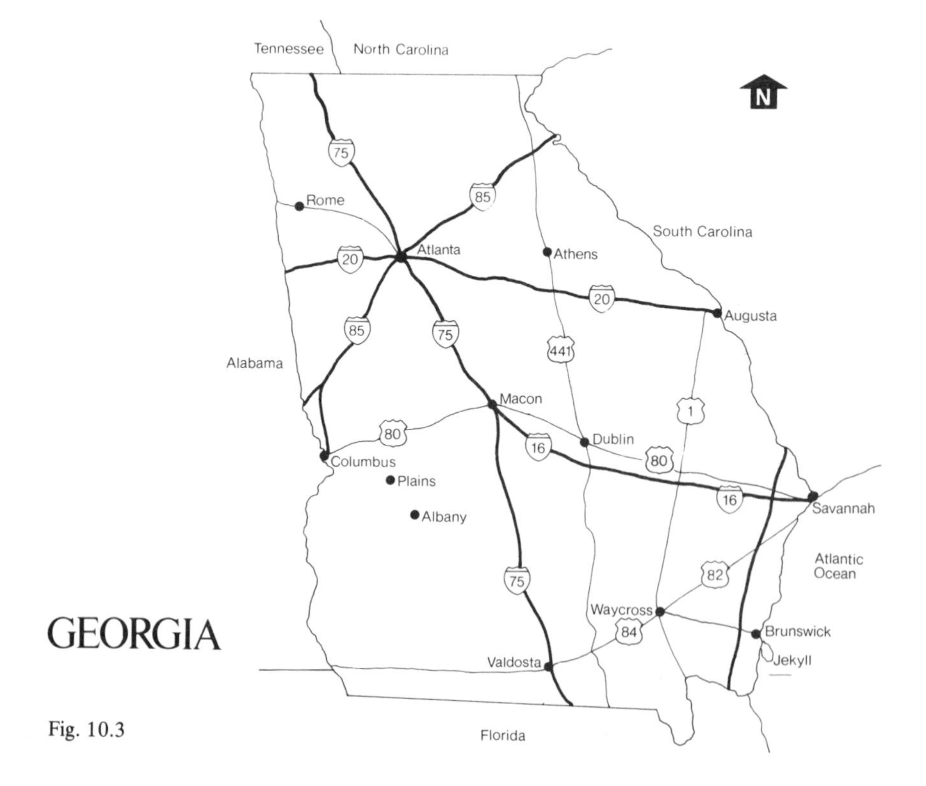

Fig. 10.3

use of addition skills, focuses on row-column coordination, and affords a second opportunity to use ratio and proportion in finding distance.

The next level of study is the national map. Emphasis here is on recognizing the systematic numbering of highways and using the key to locate capitals, historic areas, and other points of interest. Problems can be presented using time zones: "If it is 3:00 P.M. in Boston, what time is it in Los Angeles?" Rate-time-distance problems may be a challenge even in the more advanced classes: "If one averages 50 mph traveling by car, how long will it take to drive from New Orleans to Atlanta by interstate highways? If you leave New Orleans at 8:00 A.M., when will you arrive in Atlanta?"

Although a quiz should be given as the study of each map is completed, a more comprehensive unit test that covers all maps should also be given. When final exam time comes, a map of a sparsely populated state should be used to test map skills. If an atlas is not available, a hand-drawn copy of such an area can be made. It is very important that any unfamiliar map used for testing should be uncrowded and simple.

Conclusion

At least three sets of maps are needed for the unit: local, state, and national. (City and county maps are often printed together.) The same maps are used for all classes; ample time must be allowed at the end of a class period for students to refold their maps. These maps might be supplied by banks, realtors, chambers of commerce, or oil companies. Transparencies of each map, including metric overlays, should be made for the overhead projector.

In addition to improving their map-reading ability and therefore their map-skill achievement test scores, students become more aware of their hometown and its history and more knowledgeable about their state and nation. They apply graphing skills, use ratio and proportion to solve problems, and have an opportunity to measure with a ruler. But most important, they *enjoy* the unit; they often ask the next year, "Will we get to study maps this year?"

11

Graph Paper Geometry

William F. Burger

INFORMAL geometry is a vital topic in a comprehensive mathematics program in the middle school. The major components of such a comprehensive program should include the development of certain number systems (such as fractions, integers, and decimals), arithmetic operations and computation within each system, applications of each number system in various areas of study, elementary concepts in probability and statistics, extensive problem-solving experiences, the introduction of some prealgebraic concepts (e.g., variable and simple functions), *and* the development of certain topics in informal geometry. The topics highlighted in this article come in the latter category. They are (1) introduction to coordinates, (2) patterns of shapes and tessellations, and (3) symmetry and motions.

These three topics will be introduced through elementary coordinate geometry, or "graph paper geometry." The underlying structure of a coordinate system allows us to specify the exact shapes to be studied, to make neat drawings, and, perhaps most importantly, to investigate some numerical relationships among the coordinates of the figures being studied. For example, suppose we have a figure in a usual rectangular coordinate system (say with ordered pairs of the form (x, y) for the vertices) and that we multiply all the x and y coordinates by 2. We see that the figure is enlarged in a certain way. Multiplying x or y (or both) by -1 has a particular geometrical effect. So does adding to, or subtracting from, x or y (or both). Combining these arithmetical operations on the coordinates has a predictable geometrical effect, too. Thus, using graph paper geometry, we shall look for numerical, as well as geometrical, relationships within the topics to be studied.

Graph paper geometry provides a simple, yet elegant, means of introducing such topics as coordinate systems, tiling patterns, symmetry, and motions, among others. Certainly many other types of activities and instructional approaches can be used with these topics. (Several sources are listed in the Bibliography.) With informal geometry, as with many new topics in mathematics, teachers need to help children in "getting started" so that they are interested, challenged, and carefully familiarized with the concepts that they will study more deeply at some later time. The study of graph paper geometry provides students with such an initiation.

Here, then, is a sequence of activities to introduce some important topics of informal geometry in the middle school. Before they are used with the students, several things should be done:

1. Obtain a quantity of standard graph paper: either 1/4-inch paper (each small square has sides of length 1/4 inch) or 0.5-cm grid paper is good. If something other than 1/4-inch paper is used, make sure that each activity can be done in the space allowed. (Homemade graph paper can be produced with a good master and a duplicating machine. If the lines on the paper are too dark, try using the back side of the graph paper.)
2. Show the students how to draw the coordinate axes—a horizontal and vertical line that cross at a point called the *origin*. (See fig. 11.1.) Each activity will specify approximately where the origin should be on the graph paper.

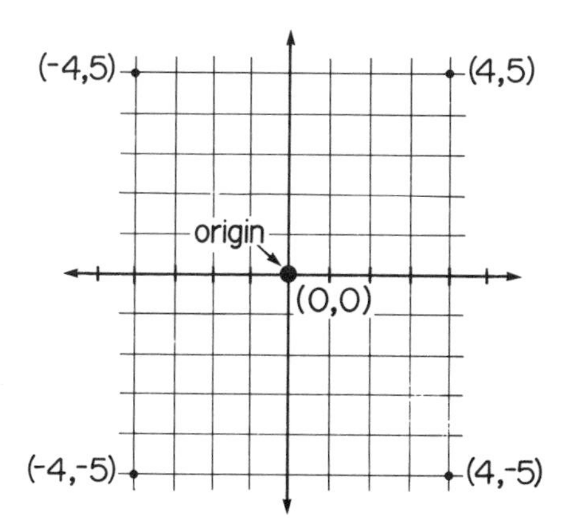

Fig. 11.1

3. Show the students how to locate points. The point where the axes cross is called the *origin* and has coordinates (0,0). To locate other points, count units to the right or left and up or down according to these rules:
   a) The first coordinate gives the horizontal distance from the origin. If the number is positive, count that many units to the right; if it is negative, count to the left.
   b) The second coordinate gives the vertical distance from the origin. If it is positive, count up that many units; if it is negative, count down.

For example, to locate (4, 5), start at the origin and count four units to the right and five up from there. (See fig. 11.1.) To locate (4, −5), count four units to the right and five down. Notice where (−4, 5) and (−4, −5) are located.

4. Be sure that students count the units correctly. Some may want to count lines or points, starting (incorrectly) with the origin as the first "point." Also, be sure that points have integers for coordinates (i.e., no fractional parts) and are located at the intersection of the lines and not somewhere inside a nearby square.

The only other equipment each student needs is a ruler with inches on one edge and millimeters on the other.

A list of the contents precedes the activities. Although it is probably better to do them all in sequence, each activity is self-contained. Solutions for activities appear at the end of each activity. (Solutions can be covered up, if desired, when photocopying activity pages for student use.)

Summary and Recommendations

Many teachers are interested in informal geometry but do not really know how to get started with it. In the middle school, students have enough familiarity with whole numbers and can easily learn the arithmetic with integers that they need, so that graph paper geometry can provide an effective context for introducing many informal geometry concepts. The activities presented here should be used specifically for that purpose—namely, to *initiate* the study of informal geometry.

Teachers will have many options once they have used these activities. Each one can be amplified as desired. It is very likely that many interesting observations and questions will be raised, and students may want to design some of their own versions of each activity. Possibly one of the topics, say tessellations, may be pursued in more generality—perhaps without using graph paper. If so, several excellent sources of additional activities are available in the Bibliography.

Contents

Introduction to coordinates

Activity	*Concepts Illustrated*	*Image(s)*
1: Maps	Locating points; making a simple map; enlarging the map	A map of Texas and an enlargement with all distances doubled
2: King Tut	Showing three dimensions in perspective; enlarging a figure	A pyramid with a square base and an enlargement of it with all distances doubled

3: The Incredible Shrinking Cube	Showing three dimensions in perspective and several enlargements of a figure	A sequence of three cubes decreasing in size by a ratio of 1/2

Patterns of shapes and tessellations

Activity	*Concepts Illustrated*	*Image(s)*
4: Diamonds Forever	Showing a repeating pattern of rhombi (diamonds)	Tessellating rhombi
5: Zigzag	Showing a repeating pattern of parallelograms; looking for number patterns	A zigzag pattern of tessellating parallelograms
6: Kites or Stingrays?	Showing a repeating pattern of kites; looking for number patterns	A tessellation with kites

Symmetry and Motions

Activity	*Concepts Illustrated*	*Image(s)*
7: Slides	Sliding (translating) a figure left/right and up/down by adding to, or subtracting from, the coordinates	A trapezoid that has been slid to four new positions
8: Flips	Flipping (reflecting) a figure over an axis by negating one of the coordinates	A trapezoid that has been flipped over the axes; a design with line symmetry with respect to the axes
9: Turns	Turning (rotating) a figure 180° by negating each coordinate	An arrow that has been turned 180°; a design (of the student's choosing) that has been turned 180°

Activity 1: Maps

1. Use the graph paper vertically (the usual position). Put the origin in the center.
2. Locate these points and label them:

$A = (1, 2)$	$G = (1, -8)$	$M = (-6, -2)$
$B = (4, 2)$	$H = (-1, -7)$	$N = (-8, 0)$
$C = (5, 1)$	$I = (-2, -3)$	$O = (-5, 0)$
$D = (5, -4)$	$J = (-3, -3)$	$P = (-5, 5)$
$E = (2, -5)$	$K = (-4, -4)$	$Q = (-2, 5)$
$F = (1, -6)$	$L = (-5, -4)$	$R = (-2, 3)$

3. Connect *ABCDEFGHIJKLMNOPQRA*. (This means *A* to *B* to *C* to *D* and so on to *R*, then back to *A*.) What state do you see?
4. Multiply the coordinates of each point by 2 to get new points A', B', C', and so on. Fill in the coordinates.

$A' = (\ 2, 4\ )$	$G' = (\ \ ,\ \ )$	$M' = (\ \ ,\ \ )$
$B' = (\ \ ,\ \ )$	$H' = (\ \ ,\ \ )$	$N' = (\ \ ,\ \ )$
$C' = (\ \ ,\ \ )$	$I' = (\ \ ,\ \ )$	$O' = (\ \ ,\ \ )$
$D' = (\ \ ,\ \ )$	$J' = (\ \ ,\ \ )$	$P' = (\ \ ,\ \ )$
$E' = (\ \ ,\ \ )$	$K' = (\ \ ,\ \ )$	$Q' = (\ \ ,\ \ )$
$F' = (\ \ ,\ \ )$	$L' = (\ \ ,\ \ )$	$R' = (\ \ ,\ \ )$

5. Locate A', B', C', and so on, and connect them in the same way as *A, B, C,* and so on.
6. What do you see?

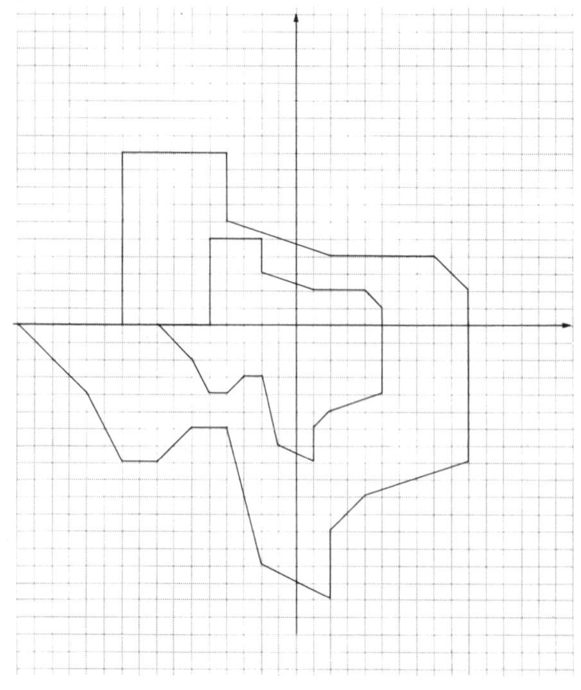

Activity 2: King Tut

1. Use the graph paper vertically. Put the origin in the center.

2. Locate these points:

   $A = (1, 5)$ $\qquad$ $D = (-4, -3)$
   $B = (7, -2)$ $\qquad$ $E = (-1, -2)$
   $C = (4, -3)$

3. Make solid lines *AB, AC, BC, CD, AD.*
   Make dashed lines *AE, DE, EB.*

4. Multiply each coordinate in *A, B, C, D, E* by 2 to get new points *A′, B′, C′, D′, E′*. Fill in the coordinates.

   $A' = (\ ,\ )$ $\qquad$ $D' = (\ ,\ )$
   $B' = (14, -4)$ $\qquad$ $E' = (\ ,\ )$
   $C' = (\ ,\ )$

5. Locate *A′, B′, C′, D′, E′*. Make solid lines A′B′, A′C′, B′C′, C′D′, A′D′. Make dashed lines *A′E′, D′E′, E′B′.*

6. What do you see?

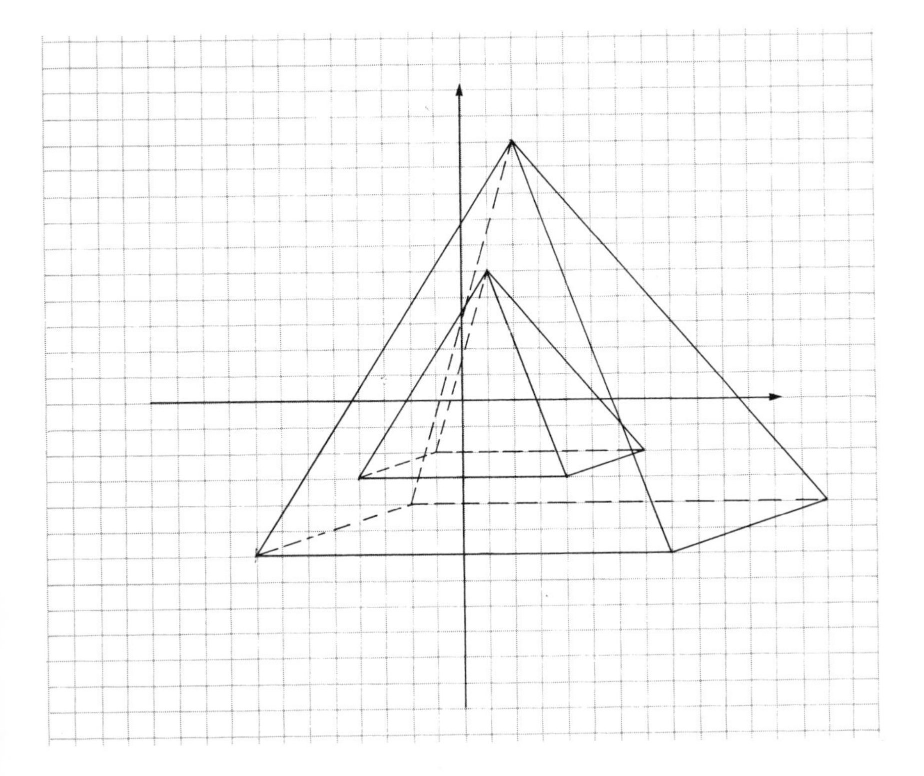

Activity 3: The Incredible Shrinking Cube

1. Use the graph paper horizontally. Put the origin in the lower left-hand corner.
2. Locate these points:

   $A = (12, 12)$ $\quad C = (20, 20)$ $\quad E = (16, 24)$ $\quad G = (24, 16)$
   $B = (12, 20)$ $\quad D = (20, 12)$ $\quad F = (24, 24)$ $\quad H = (16, 16)$

   Make solid lines *AB, AD, AH, BE, EF, EH, DG, FG, GH.*
   Make dashed lines *BC, CF, CD.* Do you see a cube?
   Which face looks closer to you, face *ABCD* or face *EFGH*?
3. Divide each coordinate in *A, B, C, D, E, F, G, H* by 2 to get new points A', B', C', D', E', F', G', H'.

   $A' = (6, 6)$ $\quad C' = (\; , \;)$ $\quad E' = (\; , \;)$ $\quad G' = (\; , \;)$
   $B' = (\; , \;)$ $\quad D' = (\; , \;)$ $\quad F' = (\; , \;)$ $\quad H' = (\; , \;)$
4. Locate A', B', C', D', E', F', G', H'. Make the same solid and dashed lines that you did in part 2. Do you see another cube?
5. Now divide each coordinate in A', B', C', D', E', F', G', H' by 2 to get points A'', B'', C'', D'', E'', F'', G'', H''.

   $A'' = (3, 3)$ $\quad C'' = (\; , \;)$ $\quad E'' = (\; , \;)$ $\quad G'' = (\; , \;)$
   $B'' = (\; , \;)$ $\quad D'' = (\; , \;)$ $\quad F'' = (\; , \;)$ $\quad H'' = (\; , \;)$
6. Locate the points in part 5. Make the same solid and dashed lines that you did in part 2.
7. Draw a straight line connecting B, B', and B''. Draw another line connecting D, D', and D''. Draw a line connecting C, C', and C''. Extend your three lines so that they cross. Do they meet at the origin? They should.
8. Turn your paper half-way around. What do you see? Do you know why this activity is called The Incredible Shrinking Cube?
9. Can you draw another cube that lines up with the three you have and is smaller than all the others? Draw it.

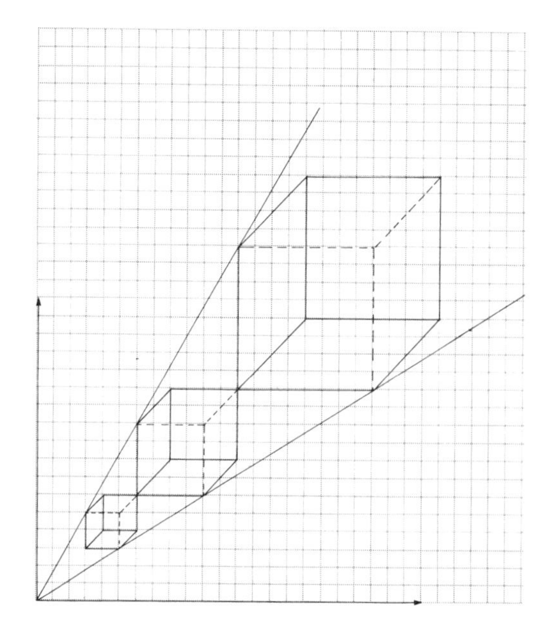

Activity 4: Diamonds Forever

1. Use the graph paper vertically. Put the origin in the center.
2. Locate these points:

$A = (0, 12)$	$F = (2, 0)$	$K = (-2, -8)$
$B = (-2, 8)$	$G = (-4, 4)$	$L = (0, -4)$
$C = (2, 8)$	$H = (-2, 0)$	$M = (2, -8)$
$D = (0, 4)$	$I = (-4, -4)$	$N = (4, -4)$
$E = (4, 4)$	$J = (-6, 0)$	$O = (6, 0)$

   Connect *ABDCA, BGHD, DFEC, GJIH, HLF, FNOE, IKL, LMN.*
3. How many small diamonds do you see?
4. Draw some more of the pattern so you have at least fifteen small diamonds.
5. Can you find some easy ways to draw more of the pattern? What are they?
6. Color the diamonds to make a pattern.

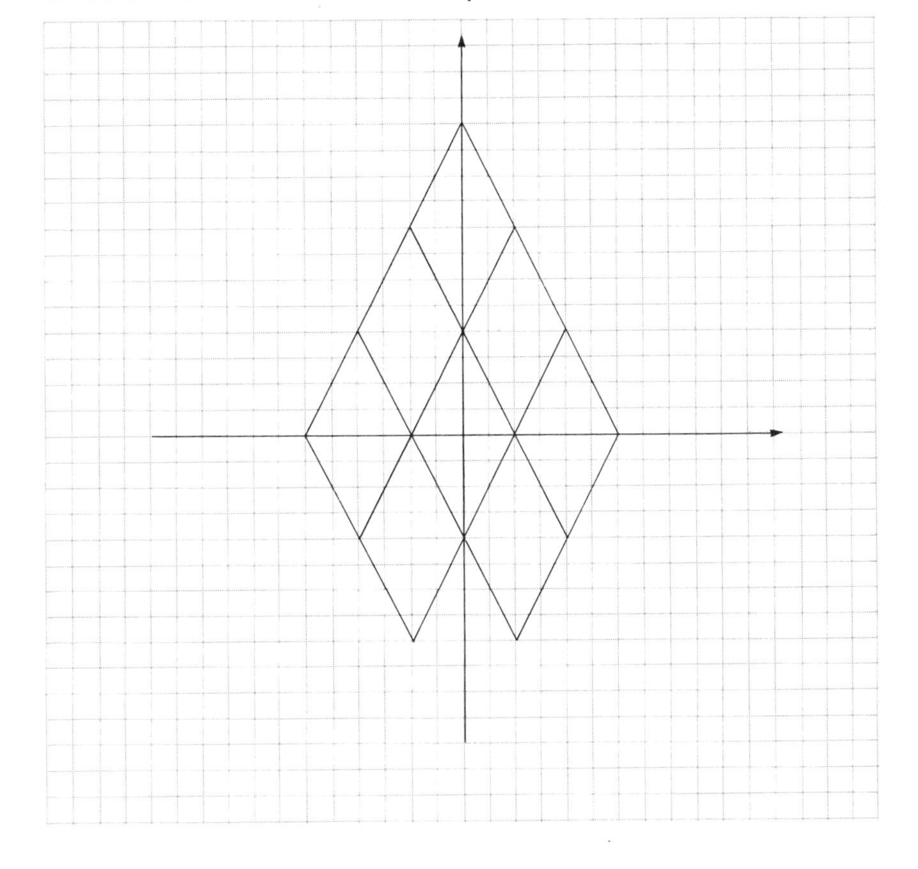

Activity 5: Zigzag

1. Use the graph paper horizontally. Put the origin in the center.
2. Locate these points:

$A = (-10, 0)$	$G = (17, 3)$	$M = (-10, 6)$	$S = (17, 9)$
$B = (-5, 0)$	$H = (12, 3)$	$N = (-5, 6)$	$T = (12, 9)$
$C = (0, 0)$	$I = (7, 3)$	$O = (0, 6)$	$U = (7, 9)$
$D = (5, 0)$	$J = (2, 3)$	$P = (5, 6)$	$V = (2, 9)$
$E = (10, 0)$	$K = (-3, 3)$	$Q = (10, 6)$	$W = (-3, 9)$
$F = (15, 0)$	$L = (-8, 3)$	$R = (15, 6)$	$X = (-8, 9)$

   Connect *ABKLA, BCJK, CDIJ, DEHI, EFGH, KNML, JON, IPO, HQP, GRQ, RSTQ, TUP, UVO, VWN, WXM.* The figure *ABKL* is called a *parallelogram.* How many parallelograms do you see?

3. Make another row of the pattern. Be sure it zigzags!
4. Look back in part 2 at the coordinates for *A, B, C, D, E, F.* Do you see any patterns in the first coordinates (−10, −5, 0, 5, 10, 15)? What do you notice about the second coordinates?
5. Look back in part 2 at the coordinates for *G, H, I, J, K, L.* Do you see any patterns?
6. List the new points that you added to the pattern in part 3. Do you see any patterns? What are they?
7. Color the parallelograms to make a pattern.

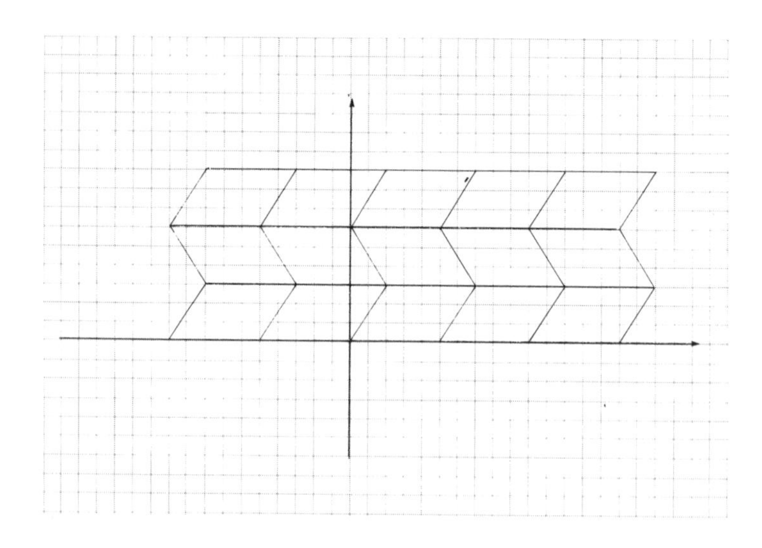

Activity 6: Kites or Stingrays?

1. Use the graph paper vertically. Put the origin in the center.
2. Locate these points:

$A = (0, 4)$	$G = (0, -4)$	$M = (0, -8)$
$B = (-2, 2)$	$H = (-2, -2)$	$N = (4, -6)$
$C = (0, 0)$	$I = (-6, 0)$	$O = (6, -4)$
$D = (4, 2)$	$J = (-8, -2)$	$P = (-2, -10)$
$E = (6, 0)$	$K = (-6, -4)$	$Q = (-6, -8)$
$F = (4, -2)$	$L = (-2, -6)$	$R = (-8, -6)$

   Connect *ABCDA, BIHC, CFED, IJKH, HGF, KRQLK, LG, GNOF, LMN, QPM.*
3. What kind of small shape do you see? Do you see more than one shape? (You should not.)
4. Draw more of the pattern, and add at least six more of the shapes.
5. Look back in part 2 at the coordinates for points *A, C, G,* and *M.* What do they have in common? Do you see any patterns?
6. Look at the coordinates for points *I, C,* and *E.* Do they have anything in common?
7. Look at the coordinates for points *B, H, L,* and *P.* What patterns do you see?
8. Do you think the shape is a kite or a stingray (or maybe something else)? Decorate it so it looks like what you think it is.
9. Color it to make a pattern.

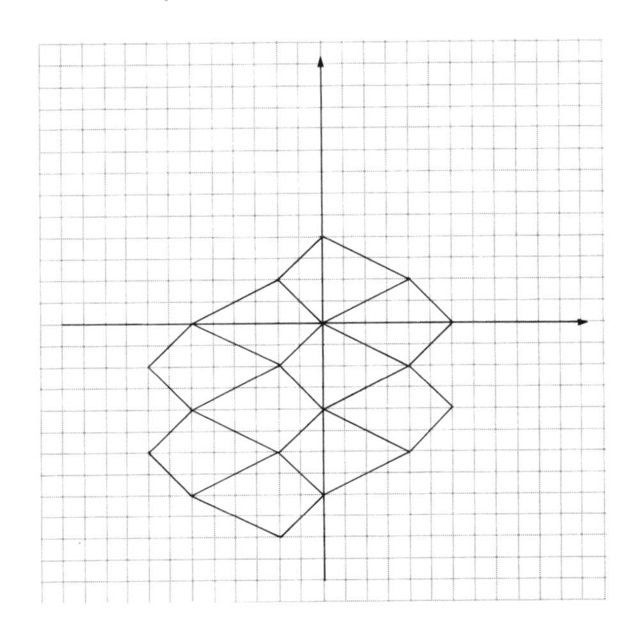

Activity 7: Slides

1. Use the graph paper vertically. Put the origin in the center.
2. Locate these points:

   $A = (-4, -2)$ $\quad$ $C = (1, 2)$
   $B = (-2, 2)$ $\quad$ $D = (5, -2)$

   Connect *ABCDA*. The figure you have is called a *trapezoid.*
3. Add 10 to each left-hand coordinate and 5 to each right-hand coordinate to get A_1, B_1, C_1, D_1.

   $A_1 = (\ 6,\ 3)$ $\quad$ $C_1 = (\ \ ,\ \ )$
   $B_1 = (\ \ ,\ \ )$ $\quad$ $D_1 = (\ \ ,\ \ )$

   Locate A_1, B_1, C_1, and D_1 and connect them to make a trapezoid.
4. Draw a straight arrow from A to A_1. How far over and how far up is it from A to A_1?
5. Add 10 to each left-hand coordinate in part 2, and subtract 5 from each right-hand coordinate. Call these new points A_2, B_2, C_2, D_2.

   $A_2 = (\ 6, -7)$ $\quad$ $C_2 = (\ \ ,\ \ )$
   $B_2 = (\ \ ,\ \ )$ $\quad$ $D_2 = (\ \ ,\ \ )$

   Locate these points and connect them to make a trapezoid.
6. Draw an arrow from A to A_2. How far over and down is it from A to A_2? What kind of motion will move the trapezoid *ABCD* onto $A_2B_2C_2D_2$?
7. What would you do to the coordinates of *A, B, C,* and *D* to slide the trapezoid *ABCD* ten squares to the left and five up? Try your guess, and locate the points.

   $A_3 = (\ \ ,\ \ )$ $\quad$ $C_3 = (\ \ ,\ \ )$
   $B_3 = (\ \ ,\ \ )$ $\quad$ $D_3 = (\ \ ,\ \ )$
8. What would you do to the coordinates of *A, B, C,* and *D* to slide the trapezoid *ABCD* ten squares to the left and five squares down? Try your guess and locate the points.

   $A_4 = (\ \ ,\ \ )$ $\quad$ $C_4 = (\ \ ,\ \ )$
   $B_4 = (\ \ ,\ \ )$ $\quad$ $D_4 = (\ \ ,\ \ )$
9. Remember that the vertices of the original trapezoid had these coordinates:

   $A = (-4, -2)$ $\quad$ $C = (1, 2)$
   $B = (-2, 2)$ $\quad$ $D = (5, -2)$

Suppose the trapezoid were slid eight squares to the right and twelve up. Without drawing it, can you give the coordinates of the vertices?

$A_5 = (\quad , \quad)$ $\qquad$ $C_5 = (\quad , \quad)$

$B_5 = (\quad , \quad)$ $\qquad$ $D_5 = (\quad , \quad)$

10. Suppose $ABCD$ were slid seven squares to the left and nine down. What would be the coordinates of the vertices? Can you find out without drawing it?

$A_6 = (\quad , \quad)$ $\qquad$ $C_6 = (\quad , \quad)$

$B_6 = (\quad , \quad)$ $\qquad$ $D_6 = (\quad , \quad)$

Now draw $A_6B_6C_6D_6$ and check your guess.

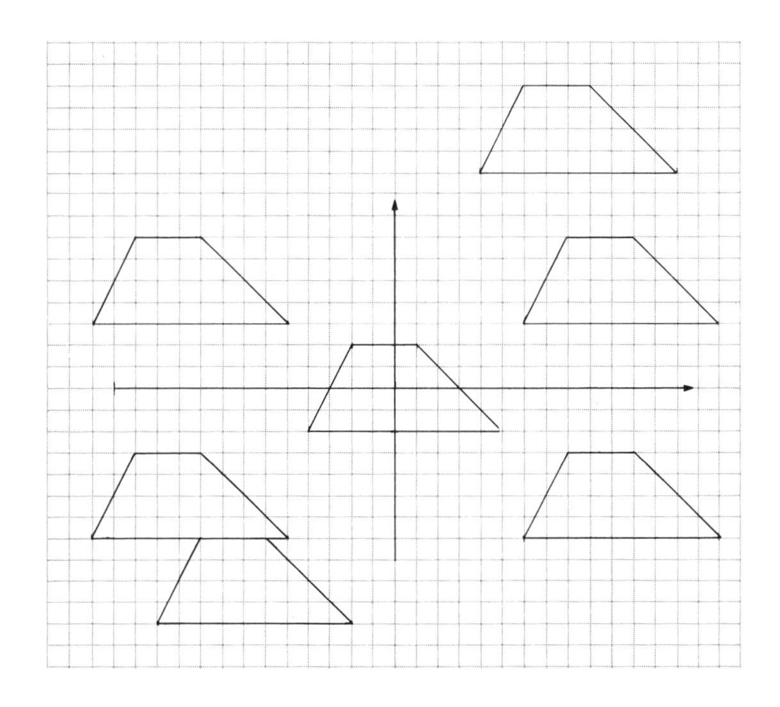

Activity 8: Flips

1. Use the graph paper vertically. Put the origin in the center. Locate these points:

   $A = (3, 3)$ $C = (8, 7)$
   $B = (5, 7)$ $D = (12, 3)$

   Connect *ABCDA* to make a trapezoid.

2. Multiply each first coordinate by -1 to get A_1, B_1, C_1, D_1. (Do not do anything to the other coordinate.)

   $A_1 = (-3, 3)$ $C_1 = (\, , \,)$
   $B_1 = (\, , \,)$ $D_1 = (\, , \,)$

   Locate these points and connect them to make a trapezoid. How is this trapezoid related to the one you made in part 1?

3. Now multiply each second coordinate in *A, B, C, D* by -1 to get new points:

   $A_2 = (3, -3)$ $C_2 = (\, , \,)$
   $B_2 = (\, , \,)$ $D_2 = (\, , \,)$

   Locate these points and connect them to make a trapezoid. How is this trapezoid related to the one you made in part 1?

4. Now take the points in part 2 and multiply each second coordinate by -1 to get new points:

   $A_3 = (-3, -3)$ $C_3 = (\, , \,)$
   $B_3 = (\, , \,)$ $D_3 = (\, , \,)$

   Locate and connect these points to make a trapezoid. How is it related to the trapezoid in part 2?

5. Take the points from part 3 and multiply each first coordinate by -1 to get new points:

   $A_4 = (\, , \,)$ $C_4 = (\, , \,)$
   $B_4 = (\, , \,)$ $D_4 = (\, , \,)$

   Locate and connect them to make a trapezoid. Is it the same as the trapezoid in part 4? It should be.

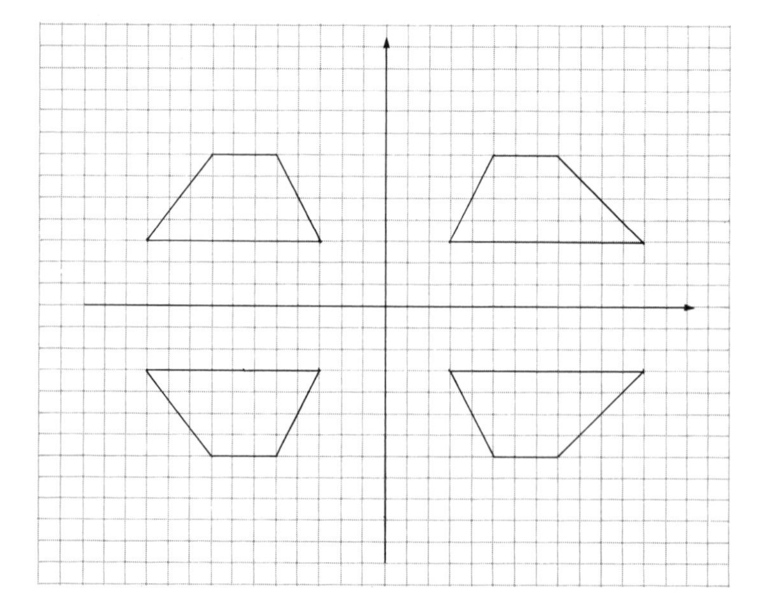

6. Start a new picture on another vertical piece of graph paper. Put the origin in the center of the page.
7. Locate these points:

   $E = (0, 0)$ $\quad G = (2, 4)$ $\quad I = (6, 4)$ $\quad K = (9, 0)$

   $F = (2, 7)$ $\quad H = (6, 8)$ $\quad J = (8, 4)$

   Connect *EFGHIJK*.
8. Multiply each first coordinate by -1 to get new points:

   $E_1 = (0, 0)$ $\quad G_1 = (\ ,\ )$ $\quad I_1 = (\ ,\ )$ $\quad K_1 = (\ ,\ )$

   $F_1 = (-2, 7)$ $\quad H_1 = (\ ,\ )$ $\quad J_1 = (\ ,\ )$

   Locate and connect. What do you see? Is your picture symmetrical? How is it symmetrical?
9. Now multiply all the second coordinates in part 7 by -1 to get new points:

   $E_2 = (0, 0)$ $\quad G_2 = (\ ,\ )$ $\quad I_2 = (\ ,\ )$ $\quad K_2 = (\ ,\ )$

   $F_2 = (2, -7)$ $\quad H_2 = (\ ,\ )$ $\quad J_2 = (\ ,\ )$

   Locate and connect. What do you see?
10. Finally, multiply all the first coordinates in part 9 by -1 to get new points:

    $E_3 = (0, 0)$ $\quad G_3 = (\ ,\ )$ $\quad I_3 = (\ ,\ )$ $\quad K_3 = (\ ,\ )$

    $F_3 = (-2, -7)$ $\quad H_3 = (\ ,\ )$ $\quad J_3 = (\ ,\ )$

    Locate and connect these points. What do you see?
11. Is your picture symmetrical? How is it symmetrical? Is the left side the mirror image of the right side? It should be. Is the top the mirror image of the bottom? It should be. Could you check this if you had a mirror?

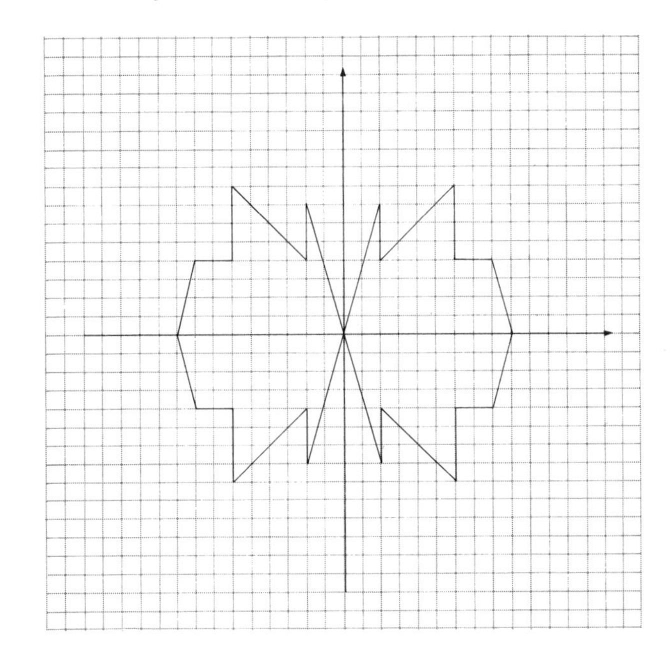

Activity 9: Turns

1. Use the graph paper vertically. Put the origin in the center.
2. Locate these points:

$A = (0, 0)$		$D = (4, 6)$
$B = (5, 10)$		$E = (1, 0)$
$C = (5, 4)$		

   Connect *ABCDEA* to make an arrow.
3. Multiply each coordinate in part 2 by -1 to get new coordinates:

$A_1 = (0, 0)$		$D_1 = (\; , \;)$
$B_1 = (-5, -10)$		$E_1 = (\; , \;)$
$C_1 = (\; , \;)$		

   Locate and connect these points. How is the new arrow related to the first one?
4. Start a new picture on another vertical piece of graph paper. Put the origin in the center.
5. Choose your own points on the graph paper. Use twelve or fewer. Record the coordinates here and name them.

   Locate and connect your points on the graph paper.
6. Multiply *each* coordinate in part 5 by -1 to get new points:

   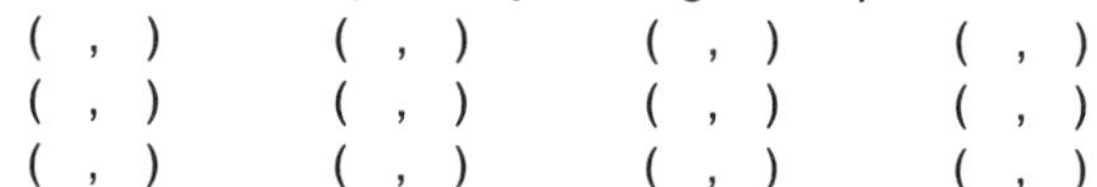

   Locate and connect your new points.
7. Look carefully at your picture. Do you see that your original design has been turned? How has it been turned?

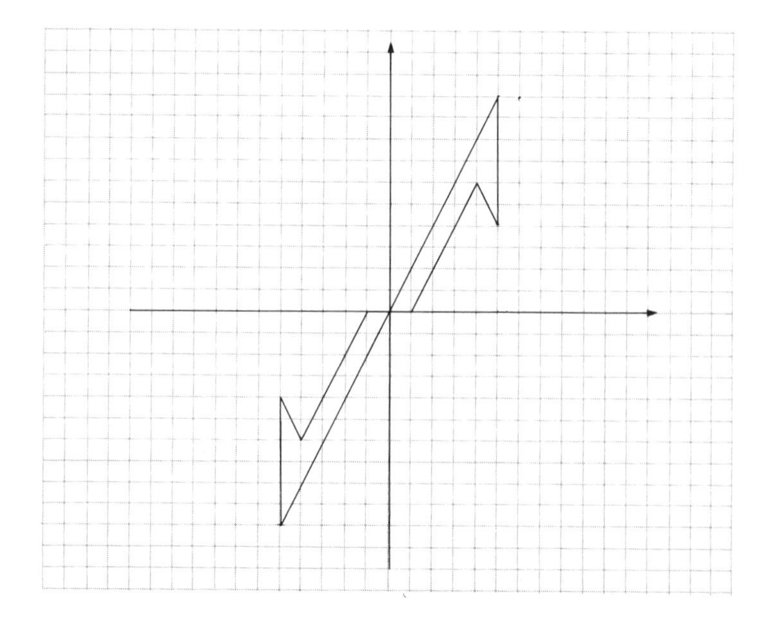

BIBLIOGRAPHY

Bezuszka, Stanley, Margaret Kenney, and Linda Silvey. *Tessellations: The Geometry of Patterns.* Palo Alto, Calif.: Creative Publications, 1977.

Boyle, Pat. *Graph Gallery.* Palo Alto, Calif.: Creative Publications, 1971.

"Geometry and Visualization." In *The Mathematics Resource Project.* Palo Alto, Calif.: Creative Publications, 1978.

Gillespie, N. J. *Mira Activities for Junior High School Geometry.* Palo Alto, Calif.: Creative Publications, 1973.

Hoffer, Alan. *Geometry: A Model of the Universe.* Reading, Mass.: Addison-Wesley Publishing Co., 1979.

Kennedy, Joe, and Diane Thomas. *Kaleidoscope Math.* Palo Alto, Calif.: Creative Publications, 1978.

O'Daffer, Phares G., and Stanley R. Clemens. *Geometry: An Investigative Approach.* Reading, Mass.: Addison-Wesley Publishing Co., 1976.

Ranucci, E. R. *Seeing Shapes.* Palo Alto, Calif.: Creative Publications, 1973.

Ranucci, E. R., and J. L. Teeters. *Creating Escher-Type Drawings.* Palo Alto, Calif.: Creative Publications, 1977.

Seymour, Dale, and Reuben Schadler. *Creative Constructions.* Palo Alto, Calif.: Creative Publications, 1974.

12

Spatial Visualization

Glenda Lappan
Mary Jean Winter

S*PACE, geometry, visualization*—what these words bring to mind for many teachers is either "We didn't have time to do those pages" or "Boys are better at those than girls." These two reactions and the attitudes that underlie them may be related in a cause-and-effect way. Fennema and Sherman (1977) found that when girls and boys were carefully matched according to the mathematics and related subjects they had studied, the differences in their performance on spatial visualization tests were not significant in two out of the four schools studied. In this same study, the correlations between mathematics achievement and spatial visualization were approximately as high as the correlations between mathematics achievement and verbal measure.

This article presents a sequence of activities developed and used with middle school students. These activities are an extension of recent work developed in a unit called "Buildings and Plans" (Lappan and Winter 1979). Success with students has been encouraging, both in their enjoyment of the activities and in their increased ability to visualize three-dimensional objects and record two-dimensional views of these objects.

In the first four activities the diagram for the models consists of a set of plans, each of which includes three grid-paper pictures (see fig. 12.3). The first picture is a plan of the *base,* or foundation, of the building, showing exactly where blocks touch the ground. The second picture is a flat "skyline" view of the building as seen from the *front;* the third is a flat skyline view from the *left-hand side.* (The skyline view can be described as what you see if you stoop until your eyes are on a level with the model.) This might seem obvious, but notice that the method for determining the base plan differs from the way in which the skylines are determined.

In order to understand the diagram, a student would have to be able to (1) see how a set of plans and its building are related to each other, (2) draw plans of an existing building, (3) construct a building from a given set of plans, and (4) construct and diagram a building and evaluate another reconstruction from the plans. These four steps, in the order given, are the guidelines for the first four activities. Activity 4 extends the work into considerations of perimeter, surface area, and volume. Activity 5 presents a new diagram drawn on isometric dot paper. This scheme encourages more mental imagery and the use of fewer visual clues. Solutions are provided at the end of article.

Activity 1: Matching

The representational scheme is introduced by exhibiting five buildings made from cubes. The buildings are placed on mats labeled A, B, C, D, and E; each building is to be matched with one of five sets of plans. Perspective drawings of the buildings are shown in figure 12.1. Notice that the paper on which each building is placed should be labeled *front* and *side* and that the building must be carefully oriented to correspond. For the skyline view, it is helpful at first to close one eye. This lessens the effect of depth. All five buildings are produced from only two distinct base plans. This forces the students to use the front and side plans to make final determinations of the correct matches.

Activity 1

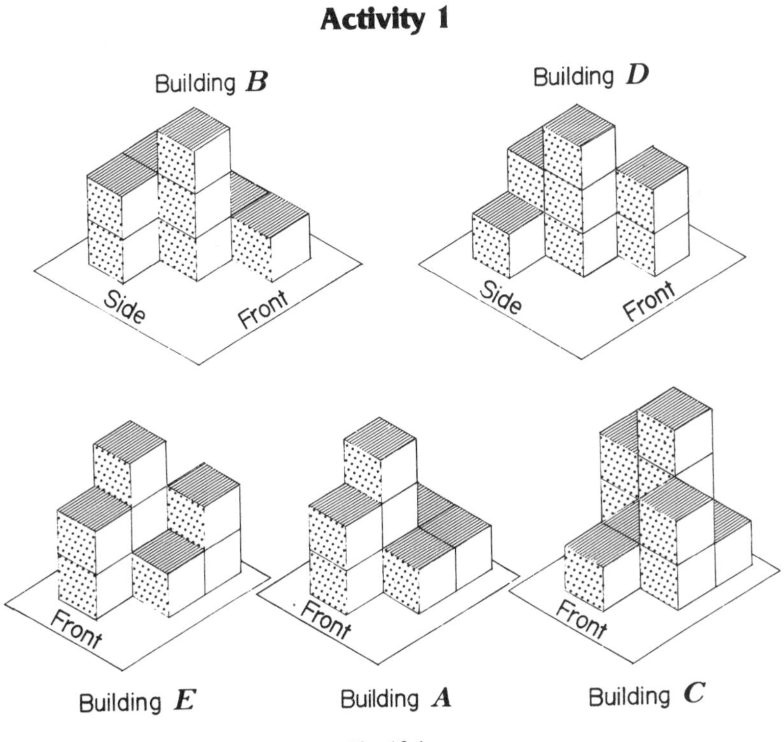

Fig. 12.1

Since these drawings do not show all the cubes in the buildings, the base plans are included (fig. 12.2). The front is at the bottom of each plan. The numbers in each square refer to the height of the building in cubes at that spot. Comparing the base plans and height numbers with the drawings in figure 12.1 will show the cubes that are hidden from view in the drawings.

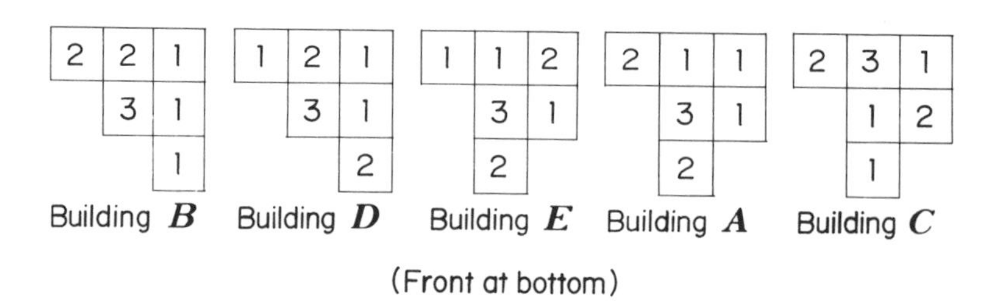

Fig. 12.2

Activity 2: Drawing plans

Buildings F, G, H, and I are set out on labeled paper so that the orientation is clear to all students (fig. 12.5). The task is to *draw* a set of plans, *base, front,* and *side,* for these buildings. Grid paper may be provided to facilitate the task. The students should be required to label their plans carefully. The major source of difficulty in this activity is that the orientation of the *base* plan must be correct. If the base plan is turned, then the plans, if used according to this scheme, will not reproduce the original building.

Since buildings G, H, and I have cubes not shown in the drawings, the floor plans with height numbers are included (see fig. 12.4). Building F is included in the activity to raise the question of whether to draw the base as a view from the top or as the bottom layer of the building. With students the ground layer base plan is usually much more helpful. That means that for our diagram □ is the correct base plan for building F.

Activity 3: Building from plans

The task in activity 3 is to construct buildings J, K, L, and M from the sets of plans given (fig. 12.7). Once the student has a building to fit the plans for J, an interesting question to ask is whether or not there are other buildings, with more or fewer cubes, that would also fit the plans. This question of nonuniqueness may arise spontaneously as different students produce correct buildings that are not identical. In fact, there are eight different realizations of building J using from eleven to fourteen cubes, as shown in figure 12.6. The other plans do produce unique buildings.

One technique that makes the process easier is to build on a sheet of paper so that the paper can be turned to view the side of the building without disturbing the building itself.

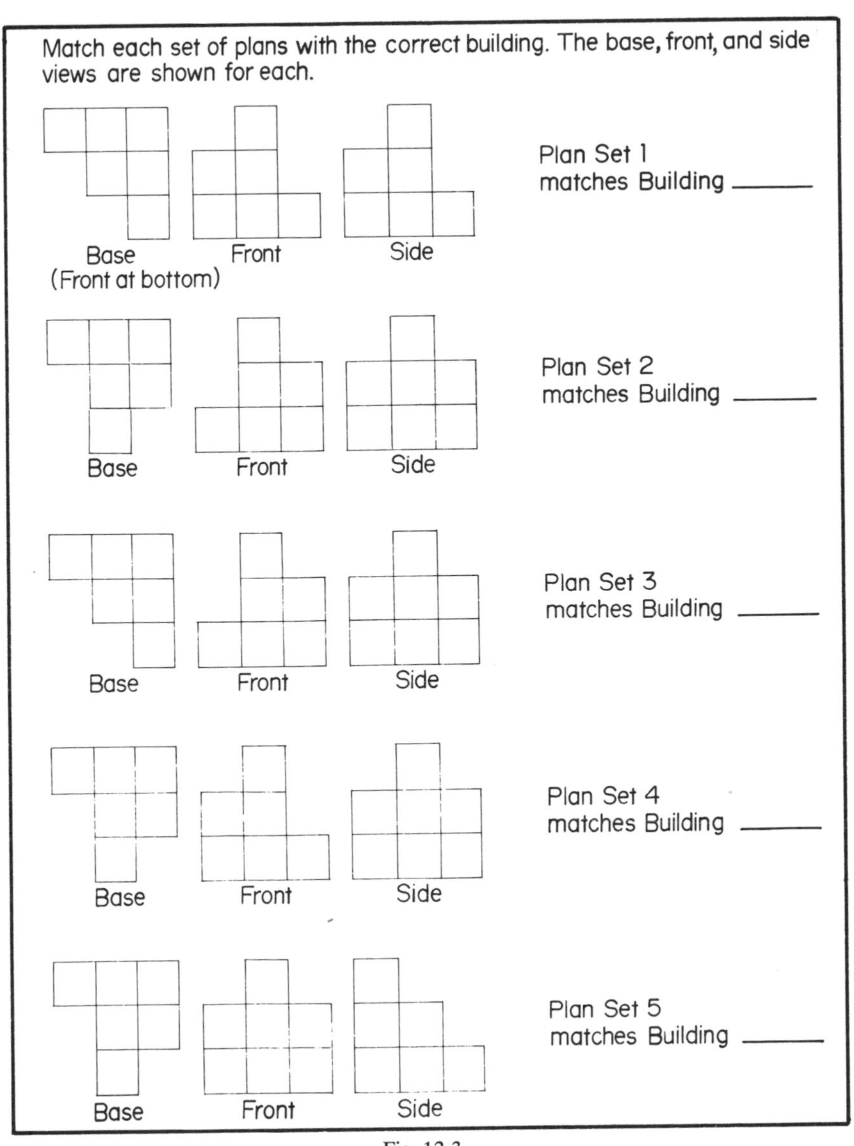

Fig. 12.3

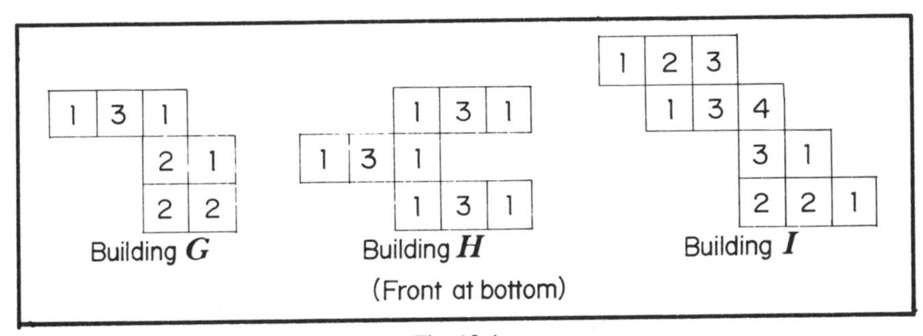

Fig. 12.4

Activity 2

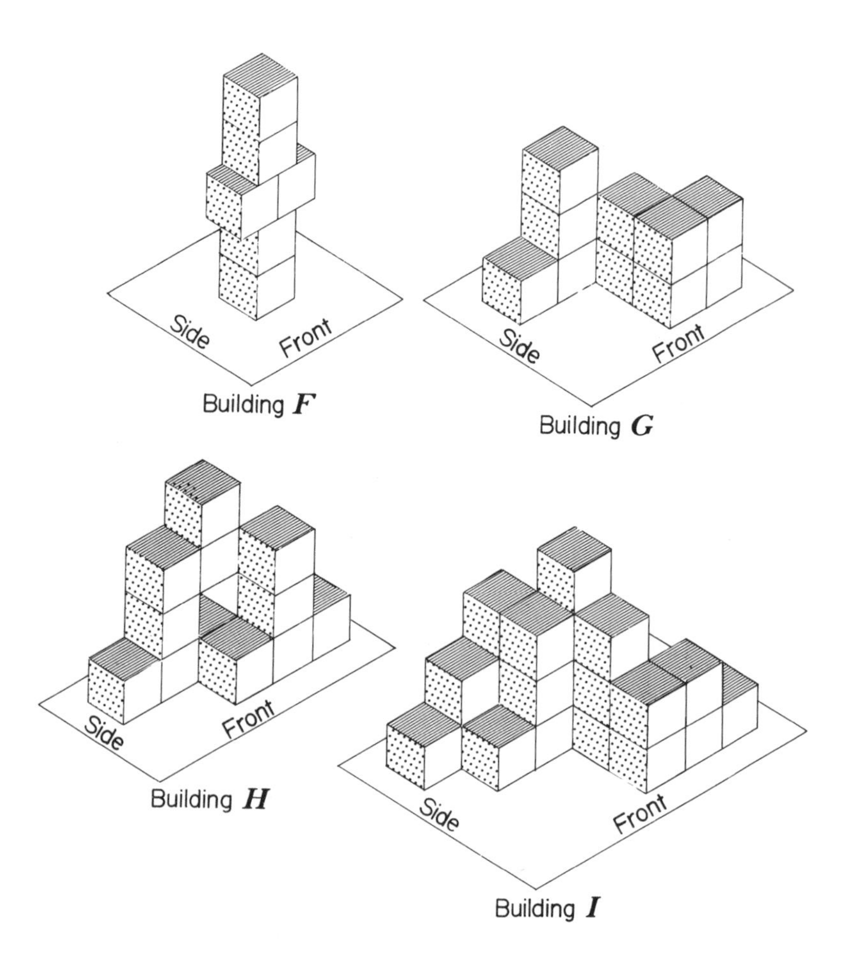

Fig. 12.5

	1 or 2		3
2	1 or 2	1	1 or 2
	2		

Fig. 12.6

Activity 3

Build each building from the given plans.

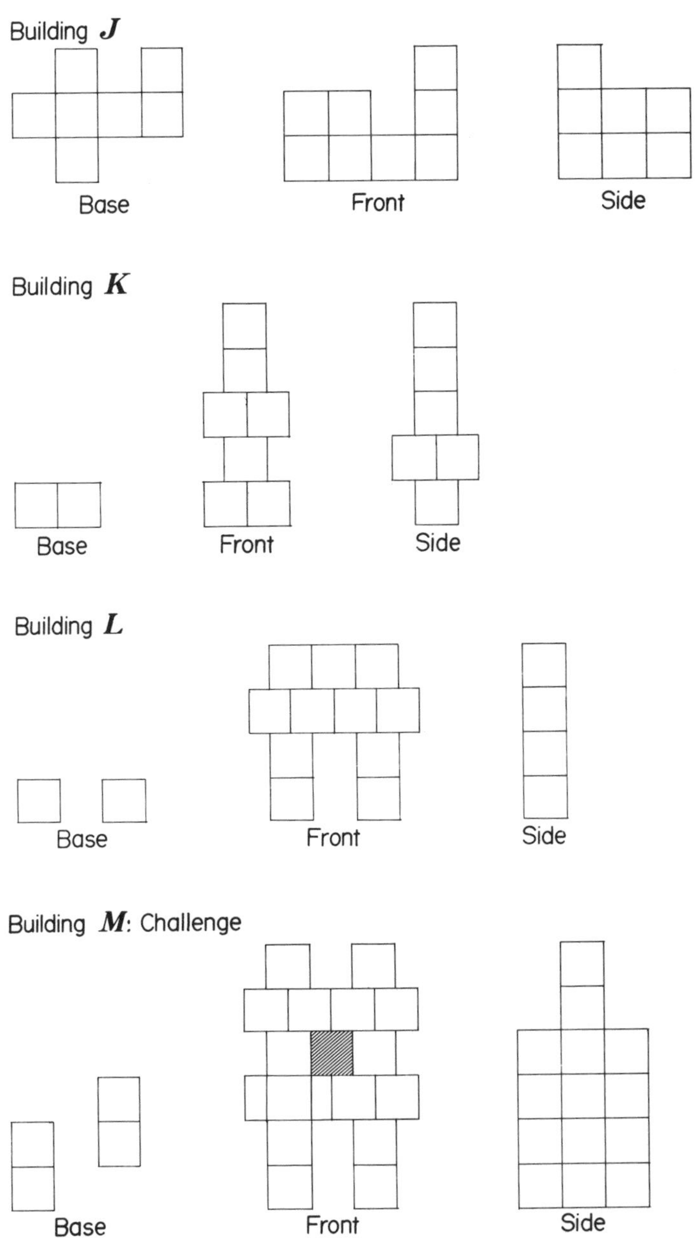

Fig. 12.7

Activity 4A: Measuring

Activity 4A can be done at any stage in the sequence of activities.

A story setting helps to extend the scheme to include the concepts of perimeter, surface area, and volume.

> An architect must specify the type and quantity of building materials as well as submit the plans for the building. The exposed surface area of each building is to be covered by square insulating panels, each of which exactly fits the face of the cubes. Around the edge of each building, a strip of drain pipe will be placed in the ground. Each pipe is as long as the edge of one cube. Specify the following numbers for each of the buildings A through E:
>
> 1. The number of blocks needed to construct the building.
> 2. The number of insulating panels needed to exactly cover the exposed surface area.
> 3. The number of pipes needed to make the strip of drain.

In the summaries with the students, make the connections between the questions and their measurement names: the number of cubes is *volume;* the number of panels, *exposed surface area;* and the length of the drain pipe, the *perimeter of the base.* The units in the story problem are cubes, square panels, and cube-edges or pipes. These correspond to cubic, square, and linear units. For buildings A through E, the data appear in table 12.1.

TABLE 12.1
DATA FOR BUILDINGS A–E

Building	Volume	Exposed Surface	Perimeter
A	10	32	12
B	10	30	12
C	10	32	12
D	10	31	12
E	10	32	12

A more difficult task is answering these same questions for buildings J, K, and L. Each realization of building J may give different answers. Here such questions as "Which version of J needs the fewest pipes? Cubes? Panels?" might be asked.

After students have completed activity 5, a variant on activity 4A is to use the isometric drawings of A through E in figure 12.5 rather than the actual models. Counting hidden surfaces and hidden cubes is tricky. In some examples the students may need the additional help of a floor plan to count correctly the volume, exposed surface area, and perimeter.

Activity 4B: Designing

The task in this activity is to design a building, draw the plans, and specify the quantities of cubes, panels, and pipes needed. (Limiting students to buildings of twelve or fewer cubes works well.) The architect's work is evaluated by having a friend build from the plans. This allows an open-ended exploration of uniqueness and nonuniqueness of plans in a setting that students really enjoy.

Activity 5: Making three-dimensional drawings

This activity introduces a different representational scheme. Isometric dot paper allows us to represent a view of a building from a corner looking down on the top. In the earlier representational scheme, the views of four people—one standing in front of the building, one behind, one on the left side, and one on the right side—could be drawn. In that scheme, the views of the persons on opposite sides of the building were simply reverse orientations. If the front view is

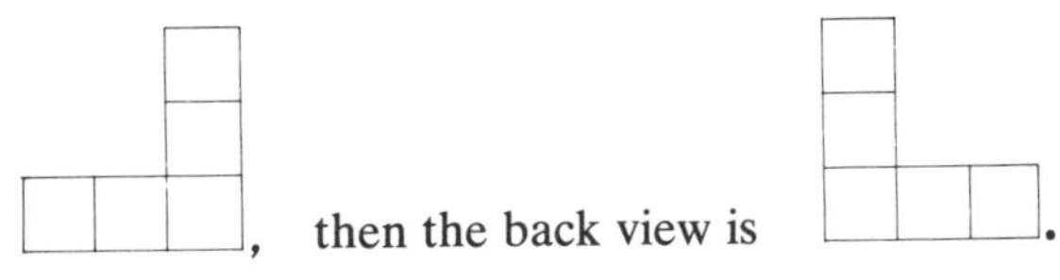

In the views on isometric dot paper a building can still be drawn from four positions—each of the four corners. However, the views from opposite corners may look quite different.

To start, give the students a piece of dot paper and challenge them to draw a small wooden cube called a "one-year-old" cube. For many middle school students this challenge requires several attempts before a correct response is drawn. Then say, "Now tip it so that I can see the bottom (or top)." This leads to the two views in figure 12.8. Now give students several of the small one-year-old cubes and ask them to build a "two-year-old" cube. After these are correctly built, ask the students to draw a two-year-old cube in both positions on the dot paper. This gives the two pictures in figure 12.9.

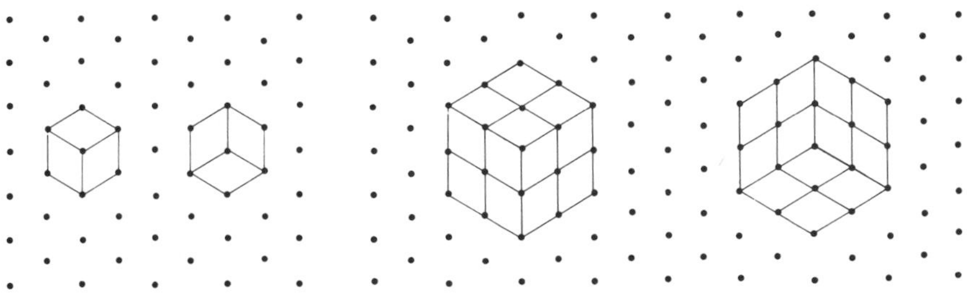

Fig. 12.8 Fig. 12.9

When the students have gotten the basic feel of the isometric dot paper by representing the "growth" of cubes, challenge them to represent a *three-long* (three one-year-old cubes joined together) in every different position that they can find. This is a marvelous challenge to observe students pursuing. At first they find one way—but they do not all find the same way. Such comments as "Mine's standing up and yours is lying down!" soon lead to the discovery of two, three, four, and finally a total of six different views. If a student is stuck at some spot, a question such as "Can you show me the bottom of this one?" will often be the only catalyst needed. Other students need the visual help of actually building and turning a three-long to find all views. When the representations are completed, the six appear as in figure 12.10.

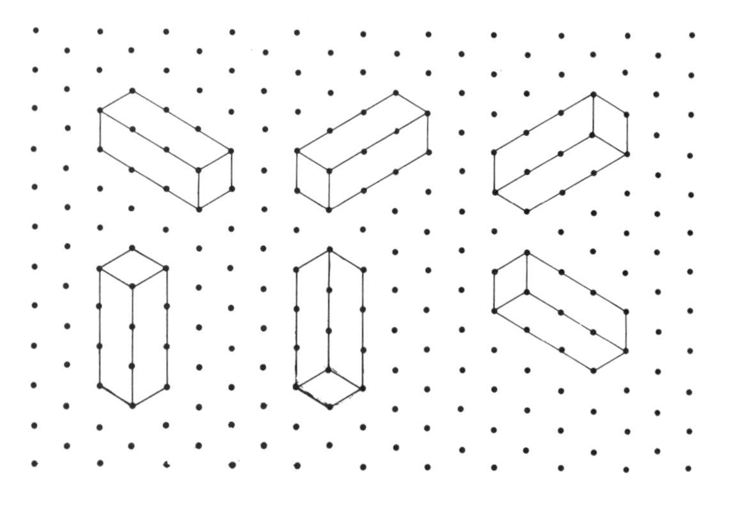

Fig. 12.10

Once the students have experienced making simple representations on dot paper, they are ready to try reading a more complicated drawing. Provide them with representations of buildings A, B, C, D, and E (from activity 1) made on isometric dot paper with no reference to orientation and no identification of the buildings. The challenge is to match each of the actual buildings with the drawing of that building and to identify the corner from which the drawing was made. The actual buildings should be displayed on pieces of paper with the corners marked 1, 2, 3, 4. This makes turning the building to view it from each corner easier.

The final step is to present the students with a simple building made from cubes and placed on a grid-paper mat with a grid the same size as the face of one of the cubes. The challenge is to make on isometric dot paper a drawing of the building from each of the four corners. The drawings should show the correct placement of the building on the mat.

As an example, figure 12.11 shows four views of a simple building that has enough different features to allow the drawings to show quite graphically the turning of the building. Note that the mat is turned also.

Four Views of Building *P*

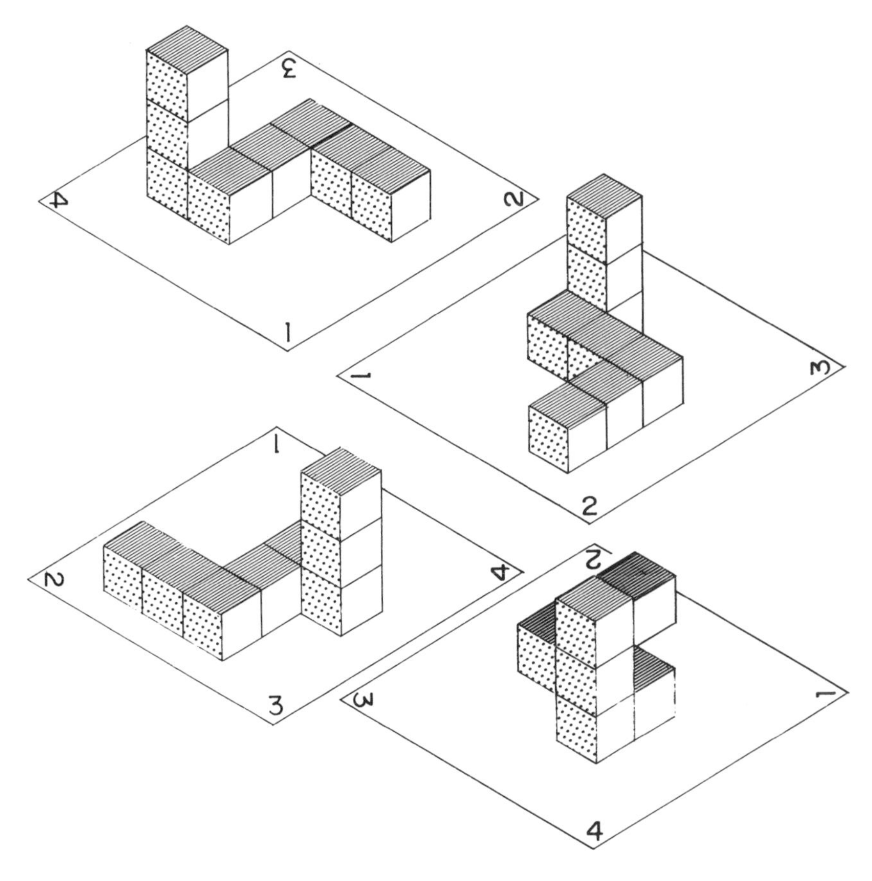

Fig. 12.11

The use of stippling to show the left faces and stripes to show the tops makes representations of this sort easier to read. Notice that in the view from corner 4, the tops of the lower level are made darker to indicate the levels.

After a few experiences making the four corner drawings of a building, some students will be able to look at a corner and mentally turn the building to draw the other three corner views. Others will need the visual stimulation of actually looking at the building from each corner in turn to produce the drawings. Throughout, however, students should be encouraged first to try to draw the views by mentally turning the building and then to check their attempts by turning the actual building and looking at each corner.

These activities provide valuable experiences in spatial visualization for students. They can also lead naturally into such extensions as similarity and scale, orientation and symmetry, counting problems, and perhaps many others that will arise from the students' explorations.

Solutions

Activity 1: Matching

Plan set 1 *B*

Plan set 2 *E*

Plan set 3 *D*

Plan set 4 *A*

Plan set 5 *C*

Activity 2: Drawing Plans

Building *F*

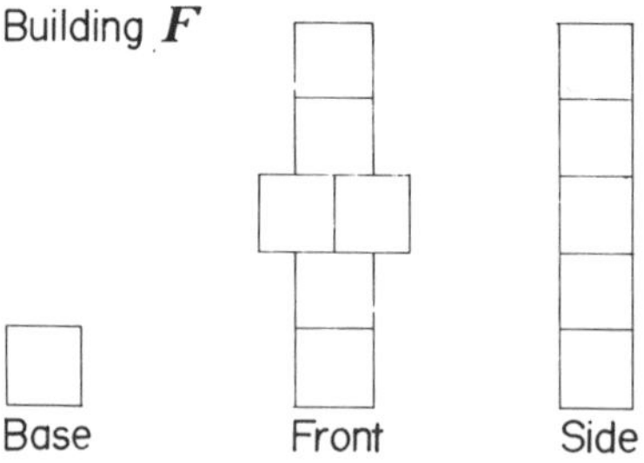

Base Front Side

Building *G*

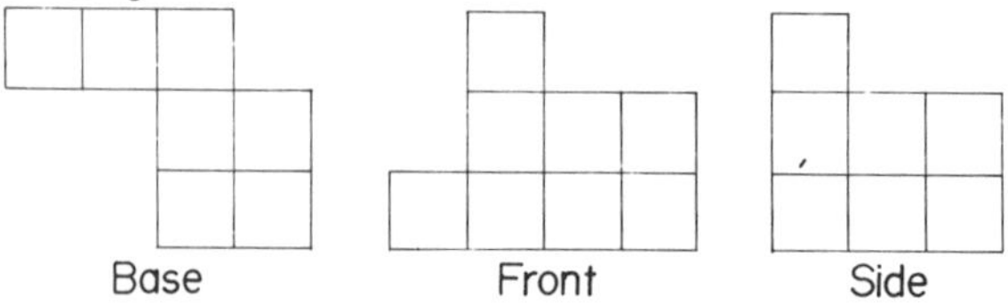

Base Front Side

Building *H*

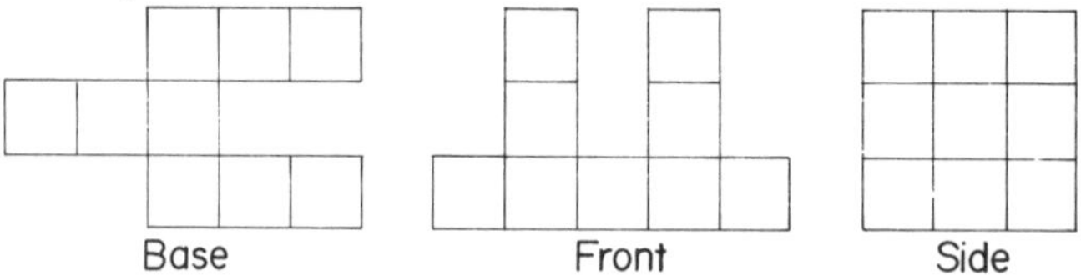

Base Front Side

Building *I*

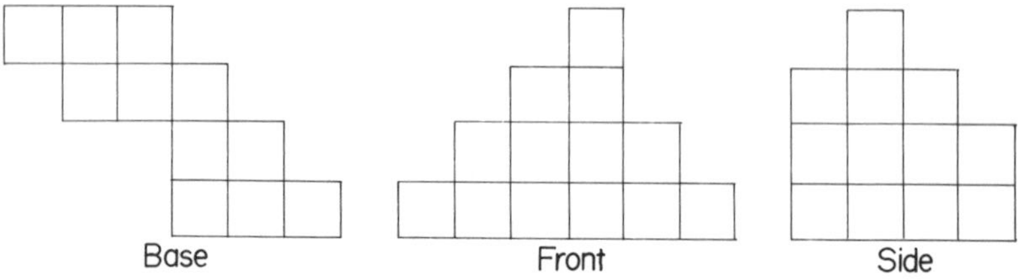

Base Front Side

Activity 3: Building from Plans

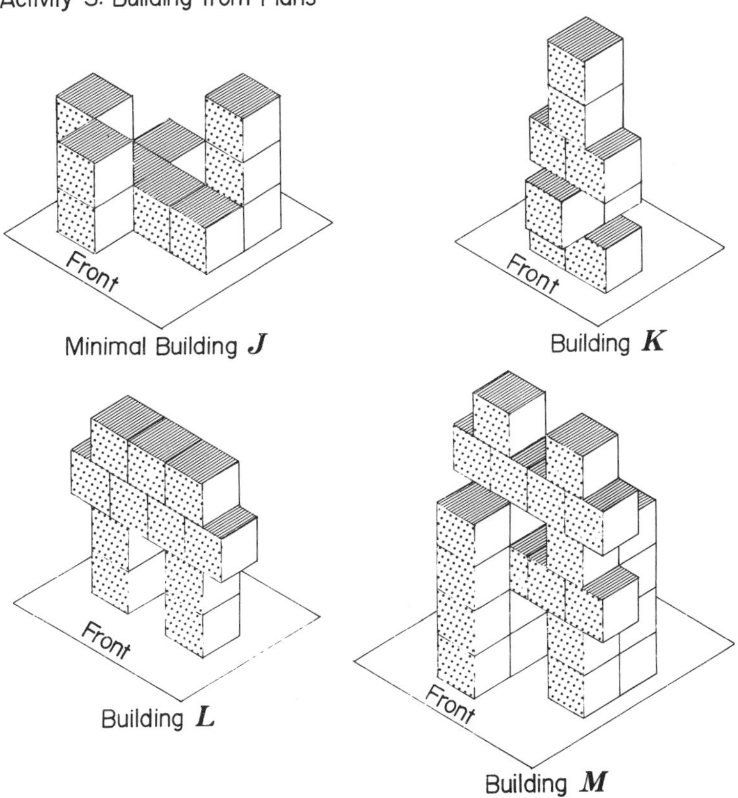

Activity 4A: Measuring

All versions of J need the same length of pipe (16 pieces). The J drawn uses the fewest number of cubes, 11. The J drawn requires the minimal number of panels, 39. The building (at right) also requires 39 panels.

	1		3
2	2	1	1
	2		

REFERENCES

Fennema, Elizabeth, and Julia Sherman. "Sex-related Differences in Mathematics Achievement, Spatial Visualization and Affective Factors." *American Educational Research Journal* 14 (Winter 1977): 51–71.

Goffree, F. "Johan—a Teacher Training Freshman Studying Mathematics and Didactics." *Educational Studies in Mathematics* 8 (August 1977): 117–52.

Lappan, Glenda, and Mary Jean Winter. "Buildings and Plans." *Mathematics Teaching* 87 (June 1979): 16–19.

13

Large Numbers and the Calculator

William B. Fisher
Jim N. Jones

BELOW is a graph found in a recent newspaper. As it shows, news reports frequently use large numbers. Whether we like it or not, large numbers are an integral part of our lives. Corporations and governments deal in millions

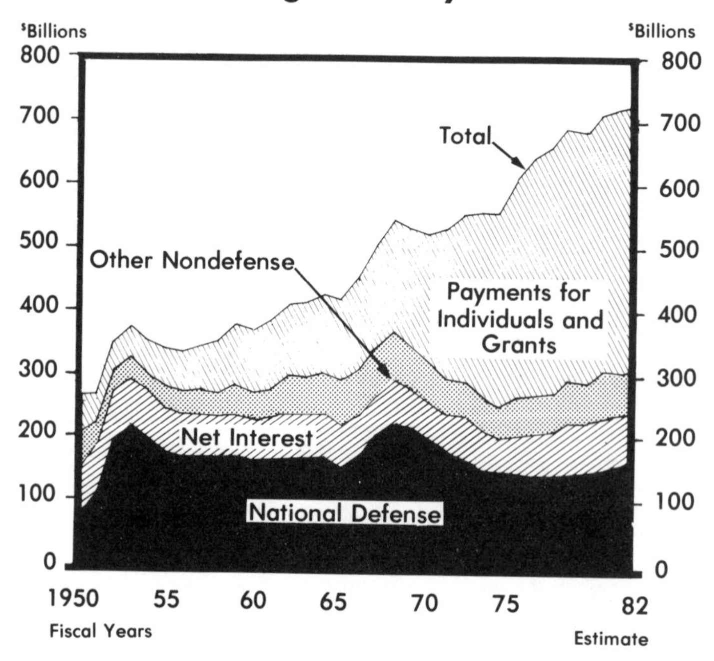

(or billions) of dollars. Americans use billions of barrels of oil each year, manufacture millions of items, and travel millions of miles (or kilometers).

Even though we are constantly subjected to large numbers, do we teach children about them? Usually not! Probably the main reason large numbers are not dealt with is that the probability of a child making an error in calculation is too high. Whereas a student can usually find the correct answer to 38 × 25, what chance does he or she have for finding the right answer to 123 456 789 × 347 821 136? The frustration and time consumed for both child and teacher are not worth it!

However, the inexpensive pocket calculator makes an investigation into large numbers a reasonable possibility, one that does not have to be frustrating. In fact, it can be very rewarding and certainly worth the time and effort. It is difficult to see how one can avoid talking and teaching about large numbers; after all, they are a part of our everyday experience:

- Did you know, for example, that in its first year of operation (1971), Disney World, near Orlando, Florida, attracted 10 700 000 visitors? Assuming attendance each year has been the same since 1971, how many people have seen this famous Mickey Mouse resort? (Based on information in the 1981 *Guinness Book of World Records.*)
- People with hay fever have already guessed this, but a single American ragweed plant can generate 8 000 000 000 pollen grains in five hours! How many grains can be produced in a field of 482 ragweed plants in a five-hour period?

Neither of these problems would probably be even looked at without a calculator because there is too much computation involved. This is not to claim that these problems are any better than the usual standard textbook problems requiring the use of a previously learned algorithm. However, children are interested in problems like these and really want to find the answers.

A summary of a series of lessons that have been presented to children from grades 4 through 9 follows. Operations with large numbers were introduced in a guided-discovery manner with good success. However, for this article only the main ideas have been presented, along with a few examples and explanations. (Answers are provided for all examples at the end of the article.) It is assumed the student has a minimal knowledge of how to use a simple four-function calculator. This presentation relies heavily on past knowledge of the usual algorithms—without this knowledge the calculator by itself will do the student little good. A nice thing about these lessons, besides their application to everyday life and the fact that one can use a simple eight-digit display calculator to work with arbitrarily large numbers, is that they reemphasize the standard algorithms. The powerful and speedy capabilities of the calculator are integrated with the student's knowledge of place value and the standard methods for adding, subtracting, multiplying, and dividing in such a way that it strengthens their mathematical abilities with all numbers.

SUBTRACTION—a Natural Starting Place

We start our adventure into the world of large numbers with the operation of subtraction.

Example 1

As captain of the spaceship *Enterprise,* you are sent to Starbase Alpha by the government of Earth to purchase one kilogram of dilithium crystals. The government provides you with a check drawn on the First Interstellar Bank of Alpha for 9 876 oogles. (The oogle is the form of currency on Alpha.) The value of 9 876 oogles is $4 389 289.44. However, the price of the needed crystals is only 7 391 oogles (or $3 284 856.03).

a) How much change do you expect in *oogles?*
b) How much change do you expect in *dollars?*

Example 1 obviously requires subtraction. It is left to the teacher's creativity to help students for whom this is not obvious. Students who see what must be done may wish to do the required subtraction without the aid of a calculator, and they should be encouraged to do so. Often a class will be split into three categories: those students who are fascinated with the calculator and will use it for even the simplest of problems; those who want to prove they can do it faster and better without the calculator; and a large group who, after a little familiarity with the calculator, learn to use it when it is more advantageous but otherwise use their own thinking abilities. However, the purpose is to have the student discover a means of doing the operations using a calculator, even though there are "too many digits." Before reading on, teachers should try to discover some method of doing the required subtraction on a calculator. It is surprising how many students even in fourth grade find a method on their own to solve the problems, but it might help to ask some questions:

"What if the amounts involved did not include cents?"

"What is the cents difference?"

"What is the dollar difference?"

"Can the dollar difference be found using your calculator?"

"Can the cents difference be found on your calculator?"

"Can you show me how the original subtraction can be done in two steps using a calculator?"

A summary of what the students discover in solving example 1 should be given and might look like the following:

Solution to example 1

Answer to (*a*): Change in oogles

```
  9 876 oogles
- 7 391 oogles
  2 485 oogles
```

Answer to (*b*): Change in dollars

```
  $4 389 289.44
-  3 284 856.03
       ??
```

To find the unknown difference, first—

- find the dollar difference only; then
- find the cents difference.

That is, separate the digits so they can conveniently be entered on your calculator.

a)
```
  $4 389 289
-  3 284 856
  $1 104 433
```

b)
```
  $ .44
-   .03
  $ .41
```

Now combine the dollars and cents, so that the answer is $1 104 433.41.

Before many complications and extensions to the process are introduced, give the students a variety of problems to practice what they have discovered. A sequence of problems that reinforce the learned process works well.

Example 2

Ms. Quackentree just inherited a bundle of money ($1 234 567.61) from her Aunt Mabel with the stipulation that she take care of Mabel's 16 cows, 3 pigs, and 14 rabbits.

Delighted with her new wealth, Ms. Quackentree wants to go on a shopping trip. However, when taxes and outstanding debts were figured on Aunt Mabel's estate, they totaled $1 024 732.75. How much money did Ms. Quackentree finally receive?

Solution to example 2

Hint: Separate the digits as follows:

$$\begin{array}{r} 1\,234\,56 \,\vdots\, 7.61 \\ -1\,024\,73 \,\vdots\, 2.75 \\ \hline \end{array}$$

Calculate:

$$a)\quad \begin{array}{r} 123456 \\ -102473 \\ \hline 20983 \end{array} \qquad b)\quad \begin{array}{r} 7.61 \\ -2.75 \\ \hline 4.86 \end{array}$$

Now combine: $\underline{\$209\ 834.86}$

Example 3

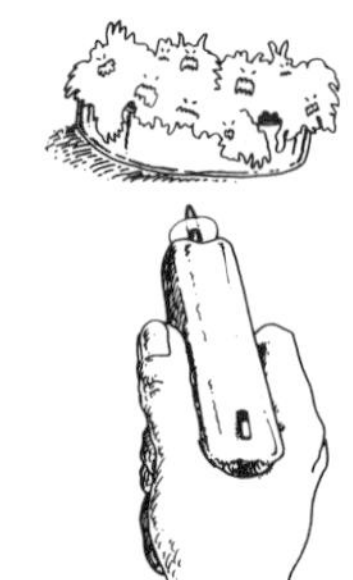

A certain colony of deadly bacteria contained 345 788 323 bacteria. One phaser beam will kill 258 796 437 of the bacteria.

a) How many bacteria are left alive after the phaser is used once?

b) How many bacteria are left after the phaser is used twice?

Sufficient problems and encouragement should be given for the student to realize that subtraction of "large" numbers can be carried out on a calculator in steps by convenient separation of the numbers into parts. Separation can occur whenever the resulting parts can be subtracted. When the teacher feels that the students are ready, they should be presented with problems that do not separate nicely and involve the process of borrowing. (See example 4.)

Example 4

Mr. Spock can compute a problem in 0.123 456 789 seconds. The computer on board the *Enterprise* requires 0.223 455 328 seconds to do the same problem. How much less time does it take Spock to solve the problem than the computer?

Solution to example 4

The difficulty in this problem is that each time the digits are separated, the subtrahend is greater than the corresponding minuend. "Borrowing" is the solution to this difficulty.

Borrow:

$$\begin{array}{r} 2\ \vdots\ \phantom{^{1}455328} \\ 0.22\not{3}\ \vdots\ {}^{1}455328 \\ -0.123\ \vdots\ \phantom{^{1}}456789 \\ \hline \end{array}$$

Separate: *a*) 0.222 − 0.123 = 0.099 *b*) 1455328 − 456789 = 998539

$$\begin{array}{r} 0.222 \\ -0.123 \\ \hline 0.099 \end{array} \qquad \begin{array}{r} 1455328 \\ -\ 456789 \\ \hline 998539 \end{array}$$

Combine (*a*) and (*b*) for the answer: 0.099998539 seconds.

If you separated this problem in a different place, did you get the same answer? You should, because our system of place value and the rules for borrowing allow you to separate a problem wherever you like.

Example 5

The budget of a certain government agency was $6 849 399.89. The agency spent $7 739 298.73. How much more was spent than was budgeted?

Once the students have mastered the process (actually, it's their traditional algorithm combined with the speed of a calculator), presenting problems using even larger numbers causes no major difficulties. (See example 6.)

Example 6

Recently the spaceship *Enterprise* was 9 992 734 567 894 011 227 kilometers from Earth. Wow! While you have been reading this, the *Enterprise*, using warp drive, sped 8 773 845 778 981 233 129 kilometers toward Earth. How far from Earth is the *Enterprise* after having traveled that distance?

$$\begin{array}{r} 9\ 992\ 734\ 567\ 894\ 011\ 227 \\ -8\ 773\ 845\ 778\ 981\ 233\ 129 \\ \hline \end{array}$$

Solution to example 6

If you guessed that some separation of digits was needed to solve the problem, you were right!

$$\begin{array}{r|r|r} 6 & 1 & \\ 9992\not{7} & 345678 & 94011227 \\ -87738 & 457789 & 81233129 \\ \hline & & \end{array}$$

a) $\begin{array}{r} 999\overset{6}{2\not{7}} \\ -87738 \\ \hline 12188 \end{array}$ *b*) $\begin{array}{r} {}^{1}345678 \\ -457789 \\ \hline 887889 \end{array}$ *c*) $\begin{array}{r} 94011227 \\ -81233129 \\ \hline 12778098 \end{array}$

Combine (*a*), (*b*), and (*c*) for the answer: <u>1 218 888 788 912 778 098 km.</u>

ADDITION—an Easy Transition

Taking enough time to get the operation of subtraction on a good foundation is worth the extra time. Addition then becomes an easy task.

Example 7

Suppose the *Enterprise*, Starbase Alpha, and the Klingons were located as shown:

How far is the *Enterprise* from the Klingon vessel?

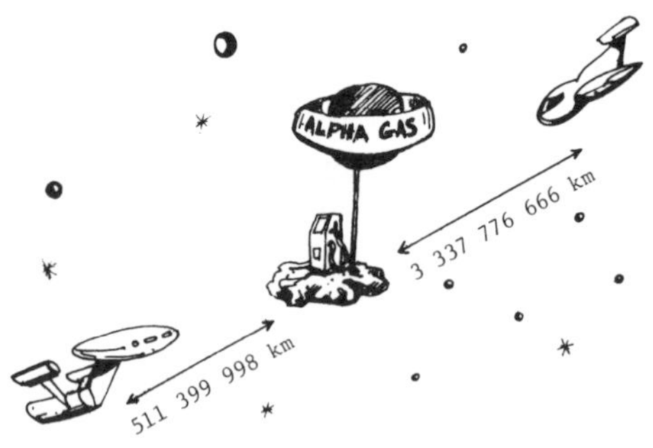

Solution to example 7

To find the distance between the *Enterprise* and the Klingon ship, you must add 511 399 998 and 3 337 776 666. These numbers have too many digits to enter into the calculator; so we take a hint from the subtraction lessons.

Separate:	333	7776666
	+ 51	1399998

Add each part individually:

a) $\begin{array}{r} 333 \\ +\ 51 \\ \hline 384 \end{array}$ *b*) $\begin{array}{r} 7776666 \\ +1399998 \\ \hline 9176664 \end{array}$

Combine the two parts, and we get the desired answer: <u>3 849 176 664 km.</u>
Caution: It isn't always that simple! For example, suppose we separate and add as follows:

a) $\begin{array}{r} 3337 \\ +\ 511 \\ \hline 3848 \end{array}$ *b*) $\begin{array}{r} 776666 \\ +399998 \\ \hline 1176664 \end{array}$

Now what do you think the answer is?

```
   3337 ¦  776666
+  511 ¦  399998
   3848 ¦ 1176664   ?  ?  (Compare with above.)
```

If you think the answer is correct, you are forgetting the important rules of "carrying" when you get more than 10. Just as

```
  48
+39
```

gives

```
 ①
  48
+39
 ①7
```

because 8 + 9 = 17, the same thing is done with larger numbers:

```
   3337     776666
+  511      399998
   3848 ① 176664
      9
```

So what you really have is

```
   3337      776666
+  511       399998
   3848     1176664
        ↘9↙
```

Combine the separate totals to obtain the answer: 3 849 176 664 km. *Look out!!!*

Whenever you get a sum for one portion of the separate parts that has *more* digits in it than the original numbers, you must "carry" the extra digit before combining the separate totals.

Example 8

A certain restaurant has sold 19 732 587 756 hamburgers and 8 773 576 895 cheeseburgers. How many burgers have they sold altogether?

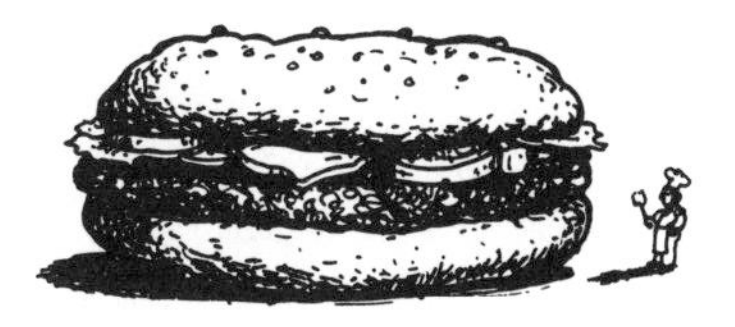

Solution to example 8

```
Separate:     19 73 ¦ 2 587 756
            +  8 77 ¦ 3 576 895
```

Try this problem, separating as follows, and see what happens:

```
   19 732 5 ¦ 87 756
+   8 773 5 ¦ 76 895
```

Can you tell where to separate these problems so you will not have to carry? (Of course, it doesn't make a difference where you separate if you obey all the rules.)

MULTIPLICATION—Knowing How Is a Necessity

By now students should see that separation is the key to everything. In fact, the secret to working with large numbers is just to separate them into "calculator size" pieces. This idea can be illustrated with multiplication.

Example 9

The population of a certain country is 225 783 671 persons. The president of this country wishes to send every citizen a 25¢ candy bar as a gift. How much will the candy bars cost the president?

Solution to example 9

How do you solve example 9? That's right, you multiply to get the correct answer! The cost of 225 783 671 candy bars at 25¢ each is 225 783 671 × 0.25. However, you cannot enter 225 783 671 in an eight-digit display calculator. Separation of the digits works as follows:

```
Separate:     225 7 ¦ 83 671
                    ¦  ×0.25
              ______¦_______
```

Calculate:

```
a)     2257            b)     83 671
      ×0.25                    ×0.25
     564.25                20 917.75
```

```
Combine:     225 7 ¦ 83 671
                   ¦  ×0.25
             ______¦_______
                 20 917.75
             56 425
             ______________
             56 445 917.75
```

If we write down all the steps, we can see clearly what took place. Note that our separation of 225 783 671 is really just an expanded form of the number: 225 783 671 = 225 700 000 + 83 671. Now when we multiply by 0.25, we are just using the distributive law:

$$\begin{aligned}225\,783\,671 \times 0.25 &= (225\,700\,000 + 83\,671) \times 0.25 \\ &= (225\,700\,000 \times 0.25) + (83\,671 \times 0.25)\end{aligned}$$

The second multiplication (83 671 × 0.25) is just problem (*b*) and can be done quickly using the calculator. The first multiplication (225 700 000 × 0.25) still has too many digits to do on the calculator, but this is easily overcome. Realizing that 225 700 000 is nothing more than 2 257 × 100 000, we can first multiply 2 257 × 0.25 (this is problem (*a*)) and then multiply the answer by 100 000. But multiplying by 100 000 just shifts the result to the left five places; that is, adds five zeros that we chose not to record in our answer.

Example 10

Suppose each candy bar in example 9 weighs 41.2 grams. How many grams of candy does the president need for his gift?

Solution to example 10

$$\begin{array}{r} 225\,783\,671 \\ \times 41.2 \\ \hline \end{array}$$

There are several ways to separate this problem:

$$a)\quad \begin{array}{r|l} 225\,7 & 83\,671 \\ & \times 41.2 \\ \hline \end{array} \qquad b)\quad \begin{array}{r|r} 225\,783\,6 & 71 \\ \times & 41.2 \\ \hline \end{array}$$

Use your calculator to find the answers for (*a*) and (*b*). Calculating separation (*a*) gives the correct answer of 9 302 287 245.2. However, separation (*b*) shows up on most inexpensive calculators as 9 302 287 225.2. Since

$$\begin{array}{r|l} 225\,783\,6 & 71 \\ \times & 41.2 \\ \hline 29 & 25.2 \\ 93022843 & \\ \hline 93022872 & 25.2 \end{array}$$

What happened? The difference between the two answers is caused by the calculator's ability to "round off" decimal numbers. In separation (*b*) the correct answer for 2 257 836 × 41.2 is 93 022 843.2. Since there are too many digits, the calculator rounded off the answer to 93 022 843! Thus, the final answer is incorrect by 20. To avoid this situation, we need to be careful about how we separate a problem. It is left to the teacher to expand this separation process to handle the multiplication of two large numbers.

DIVISION—It's Possible, Too

Division is our last operation, and once again the traditional algorithm is mimicked.

Example 11

The spaceship *Enterprise* traveled 1 234 562 030 kilometers and used 7358 liters of warp-drive fuel on a recent mission. How many kilometers per liter did the *Enterprise* average? Mr. Spock is not feeling well, and so we need your help to find the answer.

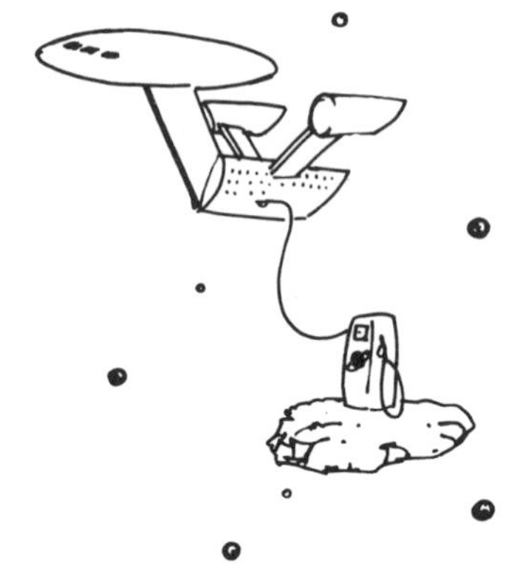

Solution to example 11

```
             1677 ¦ 85 ──► Use calculator to obtain estimates*†
7358 ) 12345620 ¦ 30
       12339366 ─────► Calculator product 7358 × 1677
             6254   30 ──► Calculator difference
             6254   30 ──► Calculator product 7358 × 85
```

$^{*}\dfrac{12345620}{7358} = 1677.8499$; so we use 1677 as our first estimate.

$^{\dagger}\dfrac{625430}{7358} = 85$; so we use 85 as our second estimate.

If your calculator has a memory function, this process can be carried out in such as way that only the digits of the answer need to be written down. However, we encourage you to write out each step of the process because the division process used in the solution is exactly the long-division process you have already learned. That is, estimate, multiply, subtract, bring down the remaining digits, and then repeat the process as needed. The only difference is that the calculator allows us to make a multidigit estimate where most of us in doing the same process "long hand" make a single-digit estimate. In fact, this observation can be made of each of the calculator processes we have discussed. They are exactly the same processes already learned except the calculator allows us to make multidigit estimates rather than single-digit estimates.

Conclusion

The students are now able to use their simple calculator to look at far more situations than ever dreamed of. As a few examples have shown, these techniques work equally well for very small numbers, too. It is hoped that their understanding of the traditional algorithms has also been increased. Teachers who try these lessons with their students should be prepared to brush up on the names for all the large place-value positions. Just the names have fascinated many students. (A simple source for these names is an ordinary Webster's dictionary under "number.") Students can now go to books, magazines, newspapers, and scientific writings in search of their own "large number" problems.

Answers to the Examples

Example 1	*a*) 2485 oogles *b*) $1 104 433.41
Example 2	$209 834.86
Example 3	*a*) 86 991 886 bacteria *b*) none
Example 4	0.099 998 539 seconds
Example 5	$889 898.84
Example 6	1 218 888 788 912 778 098 km
Example 7	3 849 176 664 km
Example 8	28 506 164 651 burgers
Example 9	$56 445 917.75
Example 10	9 302 287 245.2 g or 9 302 287.245 2 kg
Example 11	167 785 km/L

14

Decimals Deserve Distinction

Betty K. Lichtenberg
Donovan R. Lichtenberg

THE topic of decimals has been a part of the mathematics curriculum of the middle and junior high grades for a long time. Two contemporary forces that will lead to an increased emphasis on decimals as well as an earlier introduction to this type of numeral are the move in the United States and Canada toward the metric system of measurement and the ever-increasing availability of the hand-held calculator. Changing 147 centimeters to meters—a situation that may arise quite early in a child's measurement of height—will present a need for "1.47" and require some explanation. Similarly, punching [5] [÷] [8] [=] on a calculator will produce "0.625" and require explanation rather early. Society is demanding the use of decimals, and consequently we must provide sound instruction regarding their meaning.

Decimals are a logical and necessary extension of the place-value aspect of our base-ten numeration system. Procedures for computation with decimals rely on those for whole numbers. Certainly these procedures for computing with decimals are easier than some of their counterparts for common fractions. Decimal instruction can build on previous instruction in the first five or six years and can incorporate some of these ideas into a system that enables children to handle realistic problems.

The abundance of applications available for teachers can serve as a source of motivation for children. Real problems that have significant consequences involve money and percents and decimal ideas. Children are already familiar with many money concepts, and teachers can certainly take advantage of this widespread use of decimals and the complementary place-value relationships among decimals, money, and metric units. A quick glance at today's newspaper reveals the following:

- An increase in tax revenue of $2.385 billion
- A nice lunch at Red Lobster for $3.75
- Brett leads American League with .401

And so on and on. Decimals are even easier to type than fractions!

From a mathematical standpoint, decimals are equivalent to fractions that have denominators that are powers of ten. With decimals, children must think of the denominator, and the denominators are found using place-value concepts. Important mathematical objectives include computing with decimals, rounding decimals to a specified place, and ordering decimals. It is quite possible to help children gain a feeling for some of the irrational numbers. For example, students can decide that $\sqrt{5}$ is between 2.2 and 2.3 because $(2.2)^2 = 4.84$ and $(2.3)^2 = 5.29$.

Thus, as a part of the mathematics curriculum in grades 5 through 9 . . . decimals *do* deserve distinction!

Teaching for Meaning

The typical approach to decimals does not allow enough time for developing meaning, whereas inordinate amounts of time are devoted to the computational procedures. Consequently, the emphasis here is on teaching the meaning of decimals.

Models for representing decimals

Our monetary system provides a real-life application that is quite familiar to children and has a built-in emphasis on place value. Children will accept the ideas that (1) a dime is one-tenth of a dollar because a dollar is worth 10 dimes and (2) a penny is one-hundredth of a dollar because a dollar is worth 100 pennies. Then representing \$0.13 by 1 dime and 3 pennies shows that the "1" means 1 tenth and the "3" means 3 hundredths. Thus:

$$.13 = .1 + .03$$

Of course, children also know that \$0.13 can be represented by 13 pennies. This helps show why .13 can be read "thirteen hundredths."

A common error that children make is to insist that .09 > .1. With money to help them, they will be convinced that nine pennies are not worth as much as one dime, and so .09 < .1.

A model that is easy to make and use with class instruction represents decimal ideas with square regions. A master copy with several squares allows each child to participate. The fact must be established that the area of one square represents 1. Then if the square is separated into ten rows that are the same size, a row would represent 1/10, or .1 (fig. 14.1). Children can easily believe that .5 = 1/2 and that .5 > .1 (fig. 14.2).

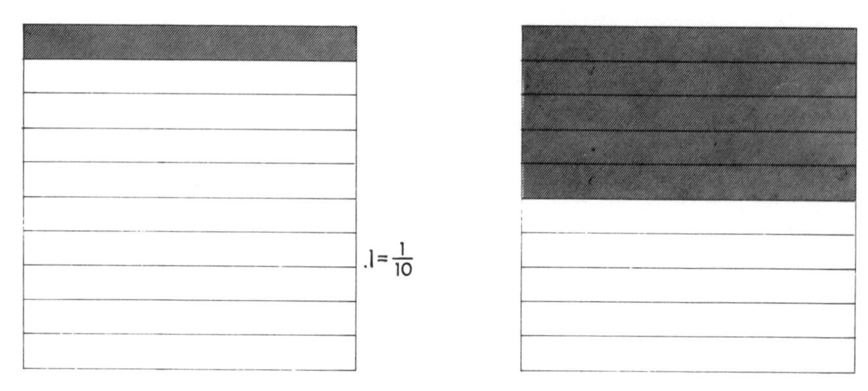

Fig. 14.1 Fig. 14.2

Now if each row of the square is separated into ten parts that are the same size, there are 100 small squares, and one of them represents .01 (fig. 14.3).

Colored pencils can be put to good use here. Using one color for tenths and another for hundredths, students can represent .13 as shown in figure 14.4. The use of color helps students see that the "1" represents tenths and the "3" represents hundredths. But they also can see that since there are thirteen small squares that are colored, the numeral .13 can be read "thirteen hundredths." It is also clear that .13 is less than .2.

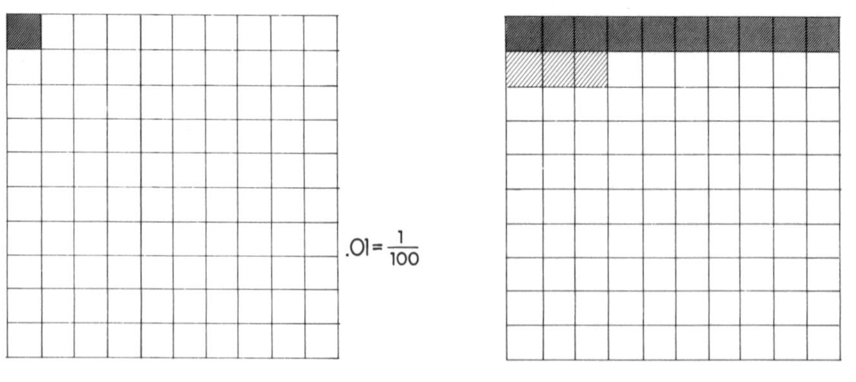

Fig. 14.3 Fig. 14.4

Teachers should be alert to the fact that the conventional way of reading decimals can be troublesome. If .3 is read "three tenths," .38 is read "thirty-eight hundredths," and .389 is read "three hundred eighty-nine thousandths," many students do not realize that the "3" means the same in each numeral. This is understandable, since it is read differently in each example. Note that this does not happen with whole numbers. For example, in each of these bracketed numerals, the "3" in the tens place is read "thirty" and the "7" in the ones place is read "seven." This would be true no matter how many digits were to the left of "37" in the numeral.

$$\left\{\begin{array}{r}37\\237\\4237\end{array}\right.$$

Once the students see that .13 can be interpreted as thirteen-hundredths, they realize the number can be represented by shading thirteen squares anywhere in the grid. They can then have fun with the activity that follows:

Color a picture that is .24 one color and .76 another.
Be creative!

Students are expected to produce designs such as those shown in figure 14.5.

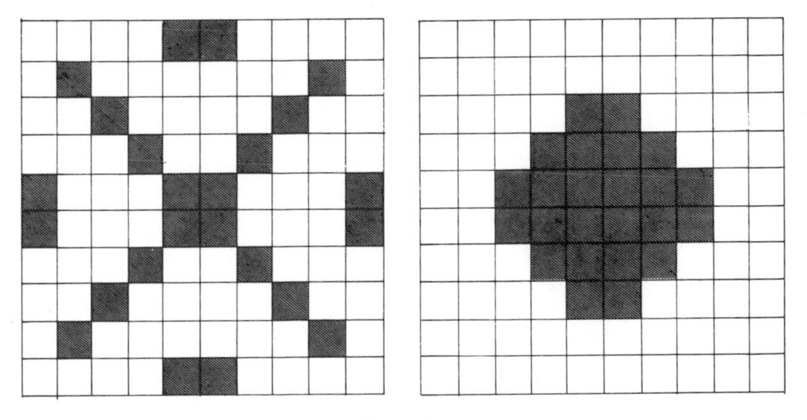

Fig. 14.5

Another model that can facilitate decimal instruction is a chain made from 100 colored plastic links (fig. 14.6). Made with only two colors, the chain could have ten blue links followed by ten yellow links, then ten blue, and so on. This results in ten separate color divisions.

Assuming that the whole chain represents 1, each link represents .01 and each different-colored segment of ten links represents .1; so, .1 = .10.

The five blue segments represent .5 and also .50. It is easy for children to see that the chain is one-half blue, and therefore they can be convinced that

$$.5 = .50 = \frac{1}{2}.$$

Substituting ten red links for one of the segments of ten yellow links offers an opportunity for additional discussion. The chain is now .1 red or .10 red,

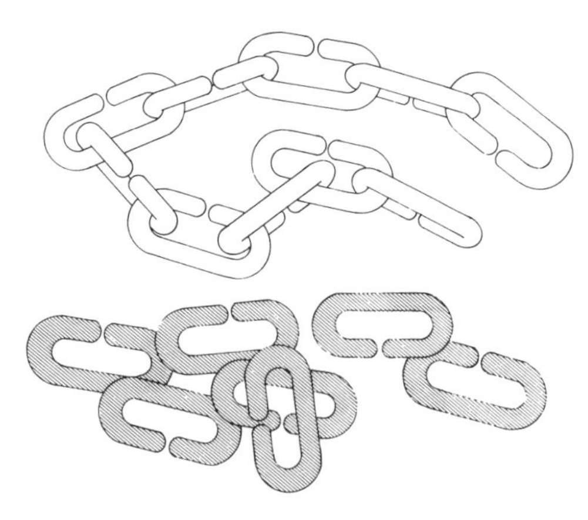

Fig. 14.6

.4 yellow or .40 yellow, and .5 blue or .50 blue. The yellow part is shorter than the blue part. So, .4 < .5. The red and yellow together are the same length as the blue part of the chain. So, .1 + .4 = .5.

The progression of decimal ideas from the concrete representation with money to areas of square regions to lengths of chains can provide a background for the more difficult transition to the number-line representation of decimals.

A segment that represents one unit is separated into ten segments of the same length (fig. 14.7). Thus, these points between 0 and 1 can be labeled .1, .2, .3, .4, .5, .6, .7, .8, and .9.

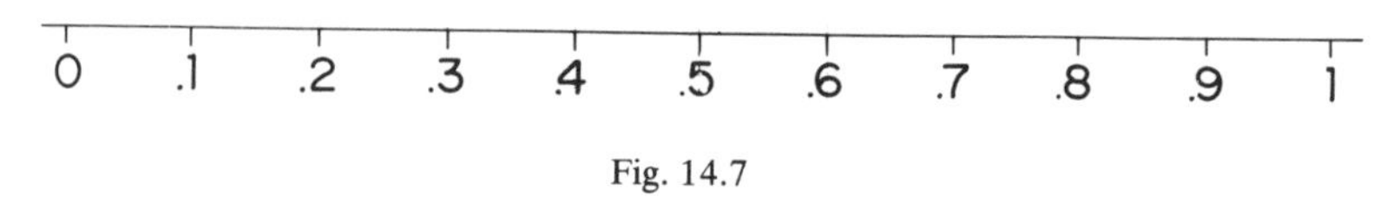

Fig. 14.7

If each of these segments is separated into ten segments of the same length, then the segment from 0 to 1 is separated into 100 parts (fig. 14.8). These points could be labeled .01, .02, .03, . . ., .99.

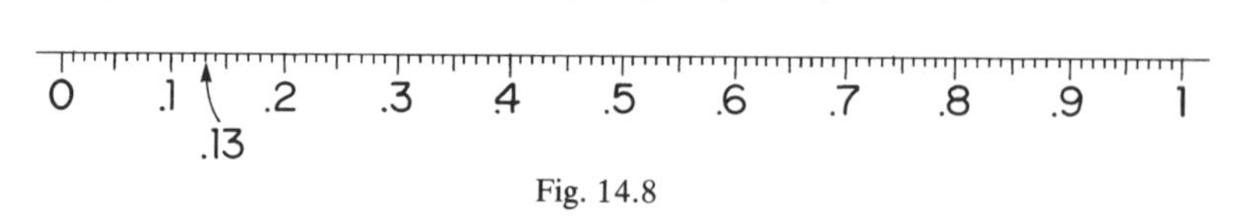

Fig. 14.8

If each of those segments is separated into ten parts, the points could be labeled .001, .002, and so on. More will be said about the number line later.

For students who have had some experience with an abacus in connection with whole numbers, this device can be used to clarify certain decimal concepts. Two abaci can be placed side by side with the explanation that the separation between them corresponds to the decimal point. (Or a single abacus can be used with a stick-on dot on the base to serve as the decimal point.) Students should see that the number 23.017, for example, would be represented as shown in figure 14.9.

Abaci help to show why we can annex zeros to the right of a decimal if we wish. For example, if asked to display .3, a student should show three beads in the tenths place (see fig. 14.10). But the display would be exactly the same for .30, and .300, and so on. So,

$$.3 = .30 = .300 = \ldots$$

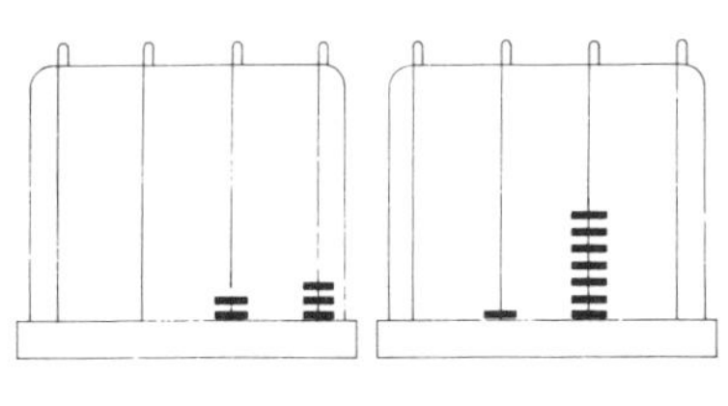

Fig. 14.9

Fig. 14.10

While using an abacus with whole numbers, students should learn that any bead can always be exchanged for ten beads in the place immediately to the right. They should see that this is true to the right of the decimal point also. This helps to show why the algorithms for addition and subtraction with decimals are essentially the same as with whole numbers.

Some hints about ordering and rounding

One of the reasons that some children find it difficult to order numbers expressed as decimals is undoubtedly due to a misconception that arises from their knowledge of whole numbers. They know that if two whole numbers differ in the number of digits, the one with more digits represents the greater number. But students must learn that this does not hold true for decimals.

Determining which of two decimals represents the greater number is a matter of scanning both decimals from left to right until one first notes a place where corresponding digits differ. A teacher can illustrate this nicely by using a device that is easily constructed from a large manila envelope and a piece of poster board. First cut off the flap of the envelope and cut out a card of poster board that will fit in the envelope with about five centimeters of it protruding (see fig. 14.11). Now write numerals on the card as shown in figure 14.12 and slide it back into the envelope.

Fig. 14.11 Fig. 14.12

The teacher holds the device in front of the class and pulls out the card as shown in figure 14.13 and asks, "Which number is larger, x or y?" Students will usually respond that they are the same. When the teacher explains that there are some more digits that can't be seen, the students will admit that they can't tell which is larger. The teacher now pulls out the card so that the hundredths digits can be seen (fig. 14.14), and the students will admit that they still can't tell.

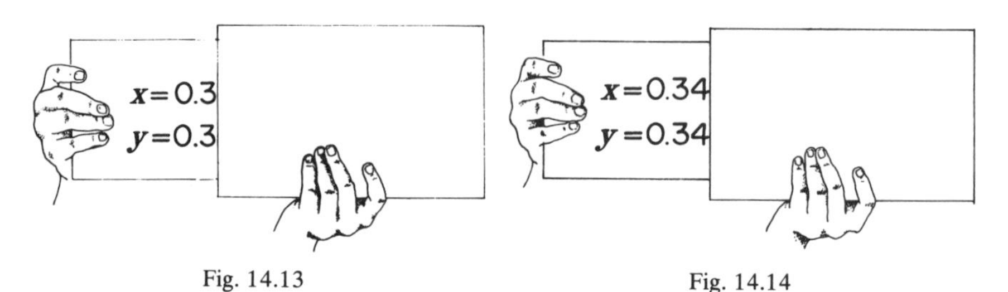

Fig. 14.13 Fig. 14.14

At the next step, when the thousandths digits are uncovered (fig. 14.15), students should agree that $y > x$. Finally, the teacher pulls the card all the way out (fig. 14.16) and makes the point that even though the numeral for x has more digits, y is still greater than x.

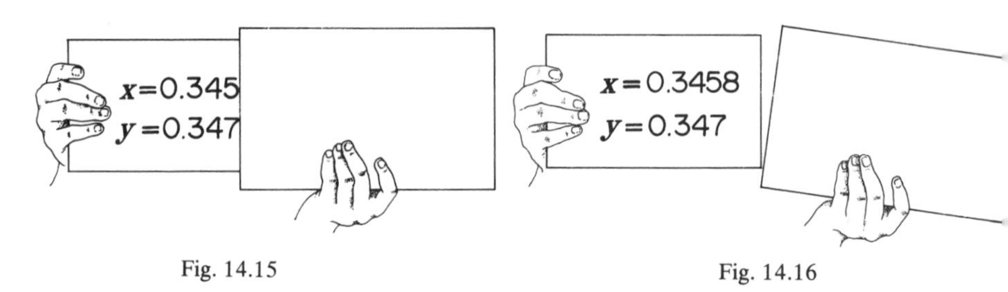

Fig. 14.15 Fig. 14.16

A device similar to the one above can be put to good use in a discussion of locating a point on a number line that corresponds to a given decimal. Again it is a matter of scanning from left to right. A teacher might begin with the display shown in figure 14.17 and say, "There are some more digits hidden in the envelope, but what can you tell me about this number now?" The best answer that students can give at this point is that the number lies between .6 and .7. However, some students are likely to say simply, "It is less than 1," or perhaps, "It is closer to 1 than to 0." Recognizing that the number is less than

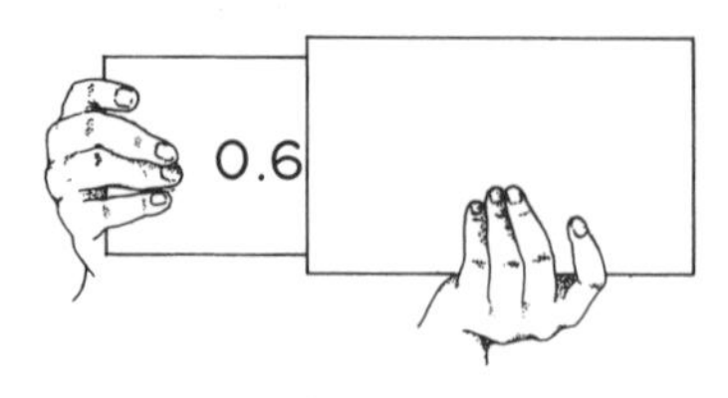

Fig. 14.17

.7 requires the knowledge that the number represented by all the unseen digits (no matter how many) is less than one tenth. When students realize this, they will agree that the number corresponds to a point on that part of the line indicated in figure 14.18.

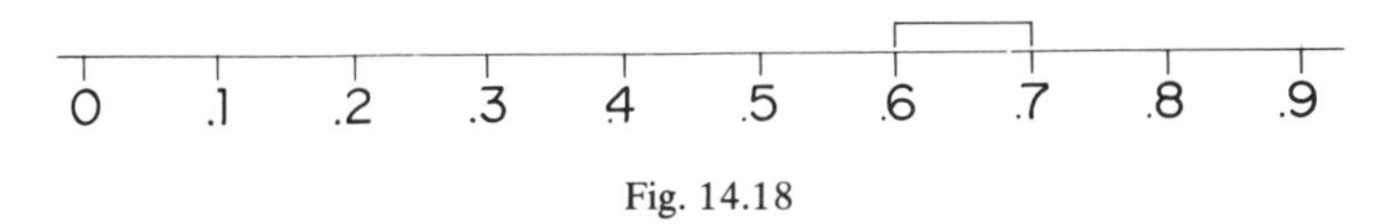

Fig. 14.18

The teacher next shows this segment magnified (fig. 14.19) and pulls out the card so that another digit is revealed (fig. 14.20). Students should now conclude that the number is between .61 and .62 and thus corresponds to a point on that part of the line marked in figure 14.21.

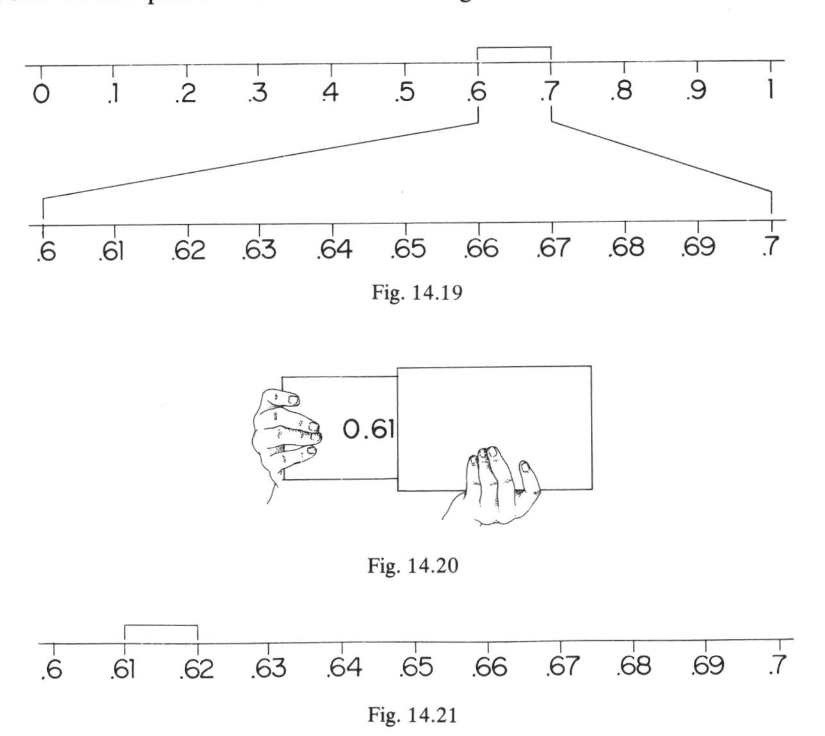

Fig. 14.19

Fig. 14.20

Fig. 14.21

Now the teacher shows this segment magnified (fig. 14.22) and pulls the card out all the way (fig. 14.23). Students can then locate the precise point (fig. 14.24) that corresponds to the number on the card.

The topic of rounding can also be worked into the discussion. When students see the display in figure 14.17 and decide that the number is between .6 and .7, the teacher can explain that "rounding to the nearest tenth" means to determine whether the given number is closer to .6 or to .7. Students will recognize that when the hundredths digit is revealed, they have

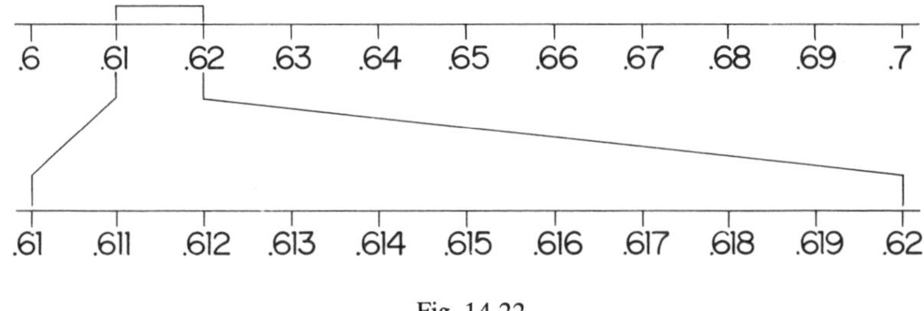

Fig. 14.22

Fig. 14.23

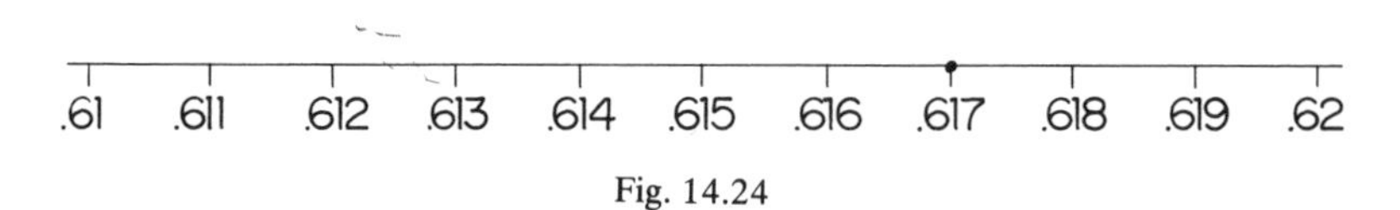

Fig. 14.24

enough information to make this decision. Figure 14.21 shows definitely that the number is closer to .6. If we are interested in rounding to the nearest hundredth, figure 14.24 shows that the number is closer to .62 than to .61.

The number line can be used to explain the conventional practice of rounding upward when there is a 5 in the next place. To illustrate, suppose that students see the display shown in figure 14.25. Assume that they have already decided that the number being revealed is between .3 and .4 (fig. 14.26). Now if there are no more digits after the 5, this number is just as close to .3 as to .4. But if it is assumed that there are some more digits, then the number is between .35 and .36 (fig. 14.27), and it is then definitely closer to

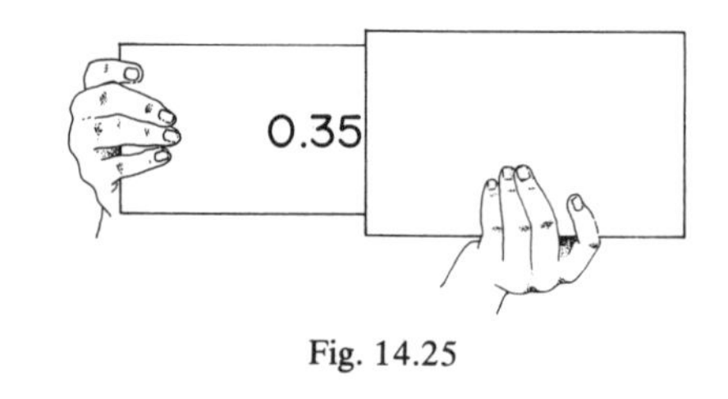

Fig. 14.25

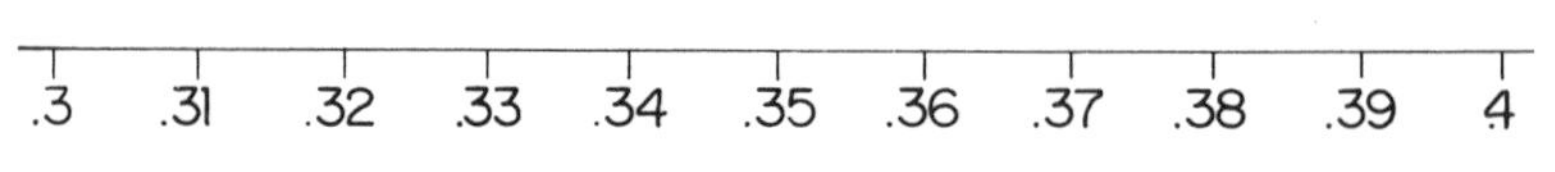

Fig. 14.26

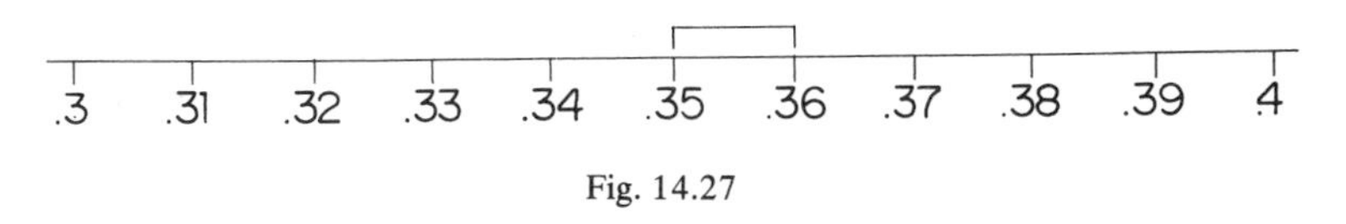

Fig. 14.27

.4. Considering the fact that decimals frequently arise in practice as the result of division that does not terminate, the assumption that there are some more digits to the right of the 5 is correct more often than not. This is the main reason for this convention.

Ideas for reinforcing

Some of the previously discussed concepts can be reinforced through the use of games or activities using index cards labeled with decimals (fig. 14.28). The directions and the decimals selected to go on the cards will vary according to the grade level and ability level of the students.

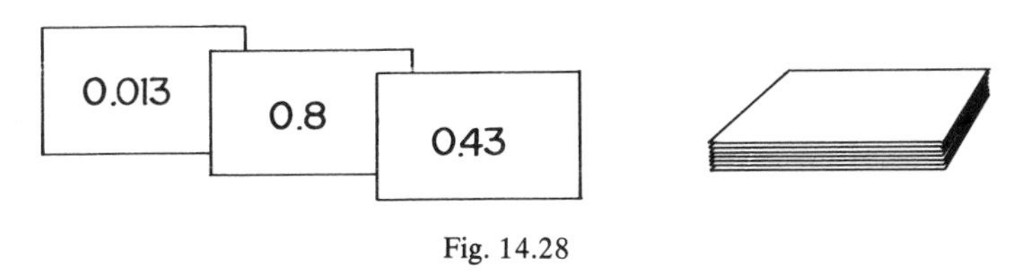

Fig. 14.28

Possible versions might include the following:

1. Have one child select a physical representation for a decimal; other children then choose the card that matches it (fig. 14.29).

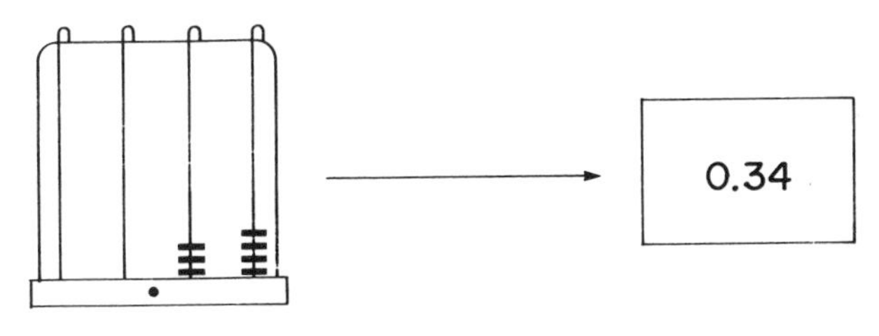

Fig. 14.29

2. Have one child select a decimal card, and then choose another child to match it with an appropriate representation (fig. 14.30).

3. Let four children play a game in which the decimal cards are dealt out, stacked face down, and then turned up one at a time. Each time four cards are played, the one who has the largest number wins all four cards. The winner is the one with the most cards when the game ends at a specified time, or when one person has captured all the cards.

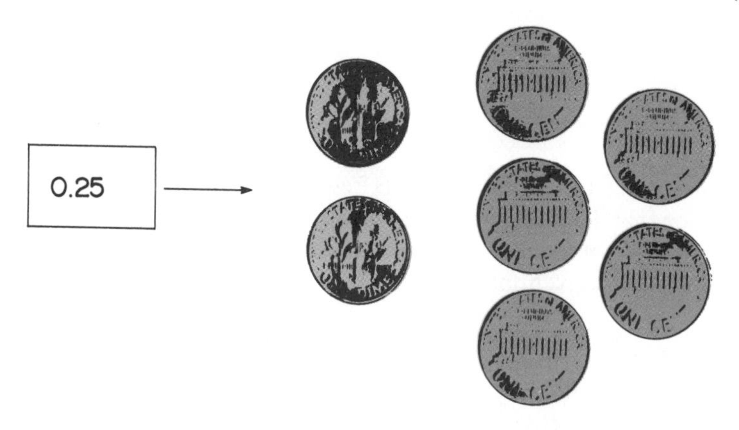

Fig. 14.30

4. Let pairs of students race at putting seven of the decimal cards in order from smallest to largest. The seven cards are dealt out face down, then turned over all at once when a signal is given. The first pair to finish says "one," the second pair, "two," and so on up to "five" to help determine winners.

5. Have a student perform a calculation (determined by the teacher) with a calculator and write the result on the board. Let other students select two decimal cards for numbers that the result falls between (fig. 14.31).

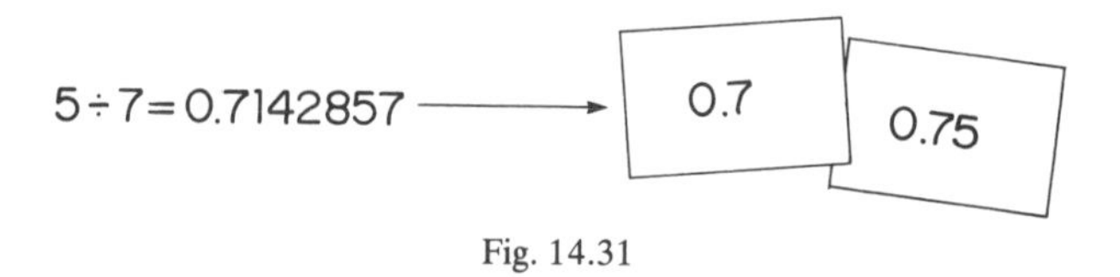

Fig. 14.31

In Conclusion

Teachers will find that time spent on developing the concepts discussed here will pay dividends in the long run. The computational processes are easier to teach and more understandable when students have a good grasp of our decimal numeration system. Furthermore, students will do a better job of interpreting decimals in the many places they occur in daily life.

15

Mathematics + Newspapers = Learning + Fun

Evelyn J. Sowell
with the collaboration of ***Rita J. Casey***

NEWSPAPERS can be used to turn on mathematics students in upper elementary and junior high school grades. The sections that follow contain sample lesson plans for teaching or reinforcing concepts, objectives for these topics and the prerequisite mathematics abilities needed, suggested preparation for instruction and activities, possible evaluation procedures, and general suggestions for using newspapers, along with some reasons you and your students will probably enjoy such lessons.

Sample Lesson 1

Topic: Arithmetical averages

Objective: The learner can compute averages of given sets of numbers.

Prerequisite mathematics abilities:

- Addition of three- and four-digit numbers
- Division with one-digit divisors

Preparation for lesson:

In the newspaper locate sets of five or six categories of numbers that are of interest to students:

- Prices for minibikes, bicycles, movie admissions, or automobiles
- Weather reports containing high or low temperatures from cities in the area
- Selected stocks from the market report
- Monthly salaries in help wanted columns

Instructional activities:

1. After you've picked out a set of five or six numbers (prices for minibikes, for example), display this information and proceed with a discussion. Ask the students what they think a typical price would be. Have them name the largest and smallest prices shown in the set, and then explain that the "typical" price will be between these two quantities. The question is how to locate the in-between, typical price. Demonstrate how to add the prices and divide by the number of prices that are added. Then show how the resulting price is between the greatest and the least quantities. You might show how each of the prices used in the computation differs from the typical price. This should help to further indicate that a typical price will lie between the highest and lowest prices.

2. Using five or six temperatures from a weather report (like the one in fig. 15.1), ask the students to point out the highest and lowest of those temperatures. Have them compute the average temperature and compare it with the highest and lowest temperatures. Give feedback about the accuracy of their computation.

U.S. TEMPERATURES			
Albuquerque	44	18	
Anchorage	46	30	0.01
Atlanta	62	30	0.13
Billings	18	14	0.02
Birmingham	59	43	0.87
Bismarck	9	2	
Boise	40	17	
Boston	41	19	
Buffalo	40	19	0.28
Charleston, W. Va.	55	35	0.26
Chicago	35	23	0.19
Cincinnati	41	26	1.10
Cleveland	38	24	0.28
Columbus, Ohio	41	28	0.92
Denver	28	7	0.05
Des Moines	22	12	0.18
Detroit	37	23	
Honolulu	81	65	
Indianapolis	42	32	0.41
Jackson, Miss.	66	44	1.34
Kansas City	24	15	0.10
Las Vegas	57	31	
Little Rock	45	34	1.30
Los Angeles	69	44	
Memphis	49	41	1.30
Miami Beach	71	68	
Milwaukee	33	25	0.19
Mpls.-St. Paul	22	10	0.20
Nashville	49	42	1.61
New Orleans	75	51	0.87
New York	41	27	
Oklahoma City	38	25	
Omaha	15	10	0.06
Philadelphia	44	18	
Phoenix	61	35	
Pittsburgh	38	21	0.58
Portland, Me.	M	M	M
Portland, Ore.	46	32	
Raleigh	58	8	
Reno	44	17	
Richmond	57	15	
St. Louis	41	24	0.53
Salt Lake City	37	29	
San Diego	66	44	
San Francisco	60	45	
Seattle	46	35	
Shreveport	51	42	0.81
Spokane	35	28	
Washington	48	25	
Wichita	28	24	

CANADIAN TEMPERATURES			
Calgary	27	10	
Edmonton	16	7	
Montreal	34	10	
Ottawa	M	M	M
Regina	9	−4	
Toronto	M	M	M
Vancouver	45	36	
Winnipeg	−2	−13	

PAN-AMERICAN TEMPERATURES			
Acapulco	86	70	
Bermuda	63	50	0.14
Bogota	M	45	0.02
Havana	81	64	
Mazatlan	77	59	
Mexico City	75	50	
Monterrey	71	55	
Nassau	75	66	T
San Juan, P.R.	87	75	0.05

Fig. 15.1. From the *Houston Post,* 2 February 1981. Used by permission.

Work several more problems with the class using sets of numbers from this or other sections of the paper, giving feedback to students on the quality of their work.

Evaluation:

Present several sets of numbers that students have not used during the instructional activity. Have students compute the average for each set.

Sample Lesson 2

Topic: Ratios

Objective: The learner can state ratios for given sets of related items.

Prerequisite mathematics ability:

- A knowledge of the meaning of fractions

Preparation for lesson:

Select sections from the newspaper that lend themselves to setting up ratios. Some suggested ratios are these:

- Number of photographs to number of news stories
- Favorite comic characters to total number of characters in comics
- Number of clothing ads to number of grocery ads
- Number of words in headlines to sections of the paper

Instructional activities:

Turn to the news section. Talk about the meaning of "news story." Then ask students to see if there is a photograph for each news story on the front page. Explain that if there were a one-to-one correspondence (one photo for each news story) this would be expressed as 1 to 1. Ask students to count the number of news stories and the number of photos on page one. Explain how to state this relationship using the term *ratio*. Use other pages of the news section to set up ratios. Then try the same kind of activity using pages from the sports or family sections. Have students point out any differences in the ratio of pictures to articles among the sections.

Use the comic section to set up ratios. Select a couple of favorite strips (see figs. 15.2 and 15.3, for example). For each strip, ask students to count the number of times one character appears in the strip they have. This is the first term of the ratio. Then have them count the total number of appearances of all the characters showing in that particular strip. This is the second term of the ratio. Work through the second strip with students, giving them feedback about the accuracy of their results.

Fig. 15.2. MOMMA by Mell Lazarus is reprinted courtesy of Mell Lazarus and Field Newspaper Syndicate.

Fig. 15.3. BLONDIE is reprinted with permission of King Features Syndicate, Inc., © 1981.

Evaluation:

Have students use other sections of the newspaper to set up ratios on their own.

Sample Lesson 3

Topic: Reading and writing names for large numbers

Objectives:

- The learner can read large amounts when they are written with numerals (1 250 000 or 317 218, for example).
- The learner can write large numbers in figures when they appear in words (three billion, $21.2 million, etc.).

Prerequisite mathematics abilities:

- Counting by hundreds to 1000
- Counting by thousands to 10 000
- Using the decimal point correctly

Preparation for lesson:

Find references to large quantities. The following are good places to look:

- News stories
- Real estate ads
- Stock market and other financial reports

Instructional activities:

Use news stories, probably from the front page of the newspaper. Ask students to circle all references to large quantities. (See figs. 15.4 and 15.5.) Review the meaning of thousands, tens of thousands, and so on. Display different ways to write each of these quantities. For instance, help them understand that "one million two hundred fifty thousand," "1 250 000," "1.25 million," and "1 1/4 million" are all the same. Let students practice reading and writing these numbers giving them feedback about their accuracy.

WASHINGTON—Staggering amounts of fraud, waste and mistakes have cost the federal government $51.8 billion to $77.3 billion, hundreds of new federal investigations and audits show.

If these losses could be stopped, the savings would all but erase the economy-crippling $55.2 billion deficit in former President Carter's 1981 budget, advisers to President Ronald Reagan acknowledge.

A Chicago Tribune study has documented the scope of the assault on the Treasury by compiling hundreds of investigative reports and audits from 16 federal agencies and other sources. Many documents were obtained under provisions of the Freedom of Information Act.

Officials such as Elmer Staats, United States comptroller general, caution that only a fraction of the fraud and waste has been uncovered.

One plank of the 1980 Republican platform was based on a Republican Study Committee analysis of 107 General Accounting Office investigations that disclosed $34 billion in waste and fraud in government spending.

Fig. 15.4. From James Coates, "Treasury Drain: Savings in Cutting Losses, Fraud, Errors Would All But Erase 1981 Budget Deficit," *Houston Post,* 1 February 1981. Used by permission.

WASHINGTON (AP)—No one can predict with a straight face that sales of queen bees to Mexico or rabbit skins to Taiwan will someday overshadow wheat, corn and soybeans at the top of the huge U.S. farm export list.

But those items—and many others even more exotic—are important to American producers and others who share in the huge world market where U.S. agricultural products dominate.

Farm exports are expected to rise to about $48.5 billion in the fiscal year that began Oct. 1, up 20 percent from $40.5 billion in 1978-80 and, for the 12th consecutive year, another record in value.

Grain and grain products, and oilseeds, mostly soybeans, and their products are expected to account for about $36.4 billion or 75 percent of this year's total export value.

In terms of quantity, those products—grains and oilseeds—will total around 136.9 million metric tons, more than 80 percent of the projected record of 170.5 million metric tons that will be exported in 1980-81.

Fig. 15.5. From "20% Farm Export Boost Seen," *Houston Post,* 1 February 1981. Used by permission.

Display data or graphs from the stock market page. Ask learners to read some of the quantities shown. Then ask them to write the numbers using numerals or words. Again, give feedback.

Evaluation:

Select quantities from sections of the paper not used in instruction, such as prices in the real estate sections, salaries in the help wanted columns, and others. Ask students to read the amounts and then to write them using a notation different from what appears in the paper.

Sample Lesson 4

Topic: Prime and composite numbers

Objective: The learner can calculate prime factorizations using division.

Prerequisite mathematics abilities:

- Knowledge of meanings for prime and composite numbers
- Division of whole numbers

Preparation for lesson:

Locate sections of the newspaper that contain two- and three-digit numbers.

Suggestions:

- Sales ads for merchandise that appeals to middle school students, such as records and tapes, sporting goods, clothing, and so on
- Selected classified ads
- Scores from sports events

Instructional activities:

Briefly review meanings of prime and composite numbers. Then direct students to a specific section of the newspaper containing at least seven to ten two- and three-digit numbers. Ask if they can spot any numbers that are prime. If they can, ask them to explain their choices. If students cannot identify prime numbers, lead them through an explanation using a couple of numbers from the newspaper. The explanation is to show how to use division to find prime factorizations.

Just to make it interesting, have the students work out the problems on the newspaper page near their set of numbers, using a marking pen or crayon. Work through a sequence of division problems, explaining as you go how you are searching for the prime factors. Have a key prepared for the exercise showing a likely sequence of division problems that present the prime factors. For example, if the number to be factored is 399, the key might look like this:

$$3\overline{)399}^{\;133} \qquad 3\overline{)133}^{\;44 \text{ r } 1} \qquad 5\overline{)133}^{\;26 \text{ r } 3} \qquad 7\overline{)133}^{\;19}$$

3 is not a factor of 133

5 is not a factor of 133

7 and 19 are prime factors of 133

$$399 = 3 \times 7 \times 19$$

Evaluation:

Give students a set of numbers from a different section of the paper. Ask them to find prime factors by using division.

Sample Lesson 5

Topic: Percentages

Objective: The learner can perform calculations involving percentages.

Prerequisite mathematics abilities:

- Multiplication and division of whole numbers
- A knowledge of the meaning of percentage and proportion

Preparation for lesson:

In the sports section, locate the standings of teams or players. Ads for merchandise that interests middle school students are also appropriate.

Instructional activities:

Help students set up and solve a proportion for one of the teams as follows:

$$\frac{\text{number of games won}}{\text{number of games played}} = \frac{\text{? percent}}{\text{100 percent}}$$

The resulting whole number percent can be expressed as a decimal fraction and compared with the percentage shown in the newspaper. Suggest that students use similar proportions for several teams and compare their results with the newspaper account of the standings. (See fig. 15.6.)

Extend the activity by asking students to figure a team's new standings after they have played additional games.

Evaluation:

Use an advertisement showing both regular and sale prices for items. Ask students to calculate what percentage of the regular price the saving represents.

NBA

Standings

EASTERN CONFERENCE				
Atlantic Division	W	L	Pct.	GB
Philadelphia	44	10	.815	—
Boston	43	10	.811	½
New York	31	22	.585	12½
Washington	26	28	.481	18
New Jersey	15	41	.268	30
Central Division				
Milwaukee	39	14	.736	—
Indiana	31	23	.574	8½
Chicago	27	27	.500	12½
Cleveland	22	32	.407	17½
Atlanta	19	34	.358	20
Detroit	13	42	.236	27
WESTERN CONFERENCE				
Midwest Division				
San Antonio	34	20	.630	—
HOUSTON	24	29	.452	9½
Kansas City	24	30	.444	10
Utah	23	32	.418	11½
Denver	20	32	.385	13
Dallas	8	45	.151	25½
Pacific Division				
Phoenix	41	16	.719	—
Los Angeles	36	18	.667	3½
Golden State	27	25	.519	11½
Portland	27	27	.500	12½
San Diego	23	30	.434	16
Seattle	21	31	.404	17½

Saturday's results

No games scheduled

Sunday's All-Star result

East 123, West 120

Monday's games

No games scheduled

Tuesday's games

San Diego at New York, 6:35 p.m.
Philadelphia at Atlanta, 6:35 p.m.
Indiana at Milwaukee, 7:30 p.m.
Kansas City at Dallas, 7:30 p.m.
Detroit at San Antonio, 7:30 p.m.
Denver at HOUSTON, 8:05 p.m.
Washington at Portland, 9:30 p.m.

Fig. 15.6. From the *Houston Post,* 1 February 1981. Used by permission.

Suggestions for Using Newspapers in Mathematics Classrooms

Newspapers used unwisely can clutter the classroom very quickly and can lead to confusion. The suggestions that follow can save time and energy and lead to more efficient student learning.

To use newspapers effectively in class, the teacher needs to plan ahead. For example, decide in advance whether all students need the same newspapers or sections and make the appropriate arrangements. If may be helpful to separate needed sections from the rest of the paper ahead of time to reduce the quantity of paper in the classroom. Decide on a definite plan for distributing the papers and for cleaning up after the lesson.

When planning, prepare several problems of the same type to keep the lesson from being a one-shot affair. You will need to know exactly where to locate information in the paper for the particular problems you want to illustrate. Probably the more diverse the sources of problems (sports, want ads, news sections, etc.), the greater the chances of appealing to the interests of individual students.

Before beginning, allow students a little time just to read the paper. Otherwise you may be involved in a mathematics lesson while the class is involved in the antics of the comic strip characters or reading "Dear Abby." Let the students know what your expectations are in using the newspaper for mathematics. Be sure your directions are clear.

Why Use the Newspaper?

Although it is true that the topics described in the sample lessons are developed in most mathematics textbooks, the newspaper can offer some advantages over other materials. There is probably no more current source of problems for mathematics than the newspaper. Each day situations change, and a whole set of new information arises from which to draw mathematics problems.

Newspapers are interesting to most students of this age. Fashions, entertainment (including sports), and ways to spend time and money literally flood the newspaper every day. Although students will probably not read all sections of the paper, most of them will take time to read about the things that are personally interesting. In addition students may see their parents or other adults significant to them reading newspapers, even when these same adults do not read other materials. This tends to encourage students to believe that newspapers are important.

Finally, the newspaper is inexpensive. Many publishers will make special arrangements to furnish papers for classroom use at minimal charges. As an alternative, students could bring papers from home. Either way, the newspaper is a low-budget, high-interest teaching material for mathematics classes.

16

Sports Card Math

Opal Kuhl

SPORTS cards can be used in the junior high school mathematics class to teach a wide variety of concepts. A more interesting approach to a topic than what students have previously experienced can help improve skills. For example, students often need to improve their skills in reading tables and graphs. The following activities and adaptations illustrate how baseball cards can be used successfully to meet the needs of both the instructor and the class.

Preparation

Begin by collecting baseball cards over the summer. Start early because by the end of September most stores have sold all the bubble gum baseball cards and are starting on football, basketball, or hockey. Select twenty or so cards for the unit. You might try to anticipate which teams will be in the playoffs and World Series and select some of those players' cards (besides some of the students' favorite players) so the cards will be more interesting to the class. Delete from the cards those figures that students will be required to calculate, and then laminate the cards. Students need an information sheet. They need to know the two types of tables used on baseball cards—batting records or pitching records. They also need (1) an explanation of the abbreviations on the cards, (2) a formula for computing batting average (hits divided by times at bat), (3) a formula for computing earned run average (earned runs divided by innings pitched times 9 [number of innings per game]), and (4) a formula for computing singles (hits minus the sum of the doubles, triples, and home runs). Finally, make up a sheet of twenty questions and give each student a copy.

The baseball card unit can be used for two or three class periods. Part of the first period is spent in explaining the information sheet. The rest of the time is used to trade the cards around so that everyone gets a chance to solve all twenty problems. The second class period and third, if necessary, are used to continue trading and solving the problems. Depending on the difficulty of the problems, the students can work either alone or in pairs.

Difficulty levels

Teachers can vary the difficulty level of the problems. At the simplest level, the cards can be used as a practice exercise in reading tables. Questions can be formulated from both the table and the other personal data on the cards. A more difficult problem is for students to use the data from the cards to make line graphs and bar graphs—for example, using the number of home runs for a certain number of years to construct a graph. Another problem could involve decimal computation and rounding. Batting averages are divided and then rounded to the nearest thousandth. Figuring a player's ERA (earned run average) involves dividing, multiplying, and rounding to the nearest hundredth. At a still higher level of difficulty, students could use the cards to study solving equations. Batting average and earned run average are computed by using two formulas that are not beyond the scope of junior high students. In fact, instead of stating the formulas for the students, the teacher could have the students themselves develop the formulas.

Using calculators

Calculators can also be used with a baseball card unit. For example, a unit designed to use only the formulas for ERA and batting average might be monotonous without calculators. Calculators could also be used in totaling lists of numbers from the tables. If the students have access to a calculator, then the number and variety of problems can be increased.

Mathematical potential

Baseball cards have great mathematical potential. Some areas in which the cards can be used successfully are reading tables, making graphs, studying statistics, using calculators, computing with whole numbers, computing with decimals, and rounding. The imaginative teacher will discover many more ways to use the cards. They can be used at various times during the school year, with different age groups, and in several topic areas.

Following are some sample problems based on the cards in figures 16.1 and 16.2. The different areas of study that these problems represent would not all necessarily be included in the same unit, and the problems serve only as a guide to the types of questions that can be asked.

- *Reading tables*

  In what year did Lou Piniella get his most home runs? (1977)

  How many years has Ron Guidry pitched for the Yankees? (5)

- *Reading tables and computing*

  How old is Ron Guidry? (Compute from birth date found on bottom line)

Fig. 16.1. Copyright 1980 Topps Chewing Gum, Inc.

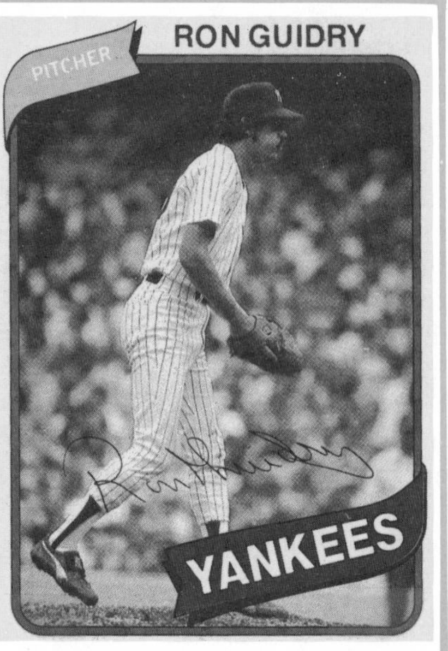

Fig. 16.2. Copyright 1980 Topps Chewing Gum, Inc.

- *Making graphs*

  Make a bar graph of Piniella's hits from 1973 through 1979.

- *Using and solving equations*

  Find Lou Piniella's batting average for 1978. (.314)
  Find Guidry's ERA for 1978. (1.74)

Cards for other sports

Other sports cards can also be used, according to the season or the interests of your class. Football cards have several types of records: tables on rushing, passing, punting, receiving, intercepting, kickoff return, or scoring. But all of them use the same format with the exception of scoring records for kickers. Each type contains four columns of data, and thus they are probably simpler to use than baseball cards. Two sample problems follow for use with football cards, based on the cards shown in figure 16.3.

- How many years did Larry Csonka play for the Dolphins? (7)
- Find Csonka's rushing average for 1975. (4.3)

Hockey cards (fig. 16.4) should also be easy to use because they contain only two lines of data—the past year's record and the lifetime record. Cards contain either playing records or goalie records. Teachers must decide which

Height: 6'3" Weight: 235 College: Syracuse
Drafted: Dolphins #1-1968 Acq: Signed as Free Agent, 1979
Birthdate: 12-25-46 Home: Lisbon, Ohio

22 LARRY CSONKA

♦RUNNING BACK

Rushing Record

YEAR	TEAM		ATT	YDS	AVG	TDS
1968	Dolphins	AFL	138	540	3.9	6
1969	Dolphins	AFL	131	566	4.3	2
1970	Dolphins	NFL	193	874	4.5	6
1971	Dolphins	NFL	195	1051	5.4	7
1972	Dolphins	NFL	213	1117	5.2	6
1973	Dolphins	NFL	219	1003	4.6	5
1974	Dolphins	NFL	197	749	3.8	9
1975	Memphis	WFL	99	421	4.3	1
1976	Giants	NFL	160	569	3.6	4
1977	Giants	NFL	133	464	3.5	1
1978	Giants	NFL	91	311	3.4	6
Lifetime Totals			1769	7665	4.3	53

Larry re-joined Dolphins as a free agent, February 22, 1979.

LARRY WAS BORN ON CHRISTMAS DAY.

©1979 TOPPS CHEWING GUM, INC. PRTD. IN U.S.A.

Fig. 16.3. Copyright 1979 Topps Chewing Gum, Inc.

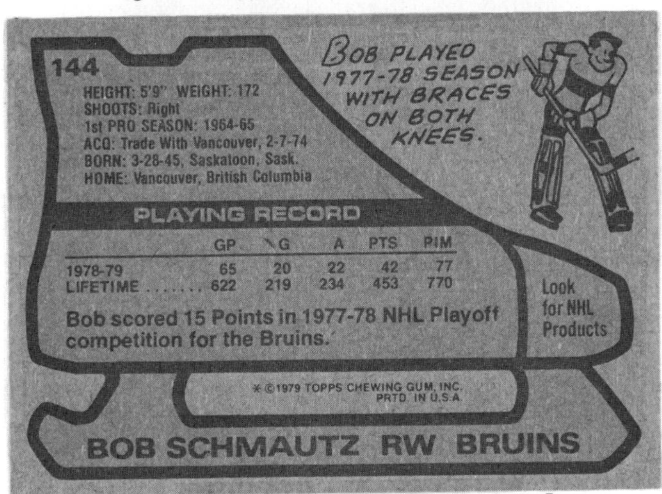

144
HEIGHT: 5'9" WEIGHT: 172
SHOOTS: Right
1st PRO SEASON: 1964-65
ACQ: Trade With Vancouver, 2-7-74
BORN: 3-28-45, Saskatoon, Sask.
HOME: Vancouver, British Columbia

BOB PLAYED 1977-78 SEASON WITH BRACES ON BOTH KNEES.

PLAYING RECORD

	GP	G	A	PTS	PIM
1978-79	65	20	22	42	77
LIFETIME	622	219	234	453	770

Bob scored 15 Points in 1977-78 NHL Playoff competition for the Bruins.

Look for NHL Products

©1979 TOPPS CHEWING GUM, INC. PRTD. IN U.S.A.

BOB SCHMAUTZ RW BRUINS

Fig. 16.4. Copyright 1979 Topps Chewing Gum, Inc.

sports interest their students and which sport cards offer the most potential for their specific class.

This article is in no way a complete presentation of how sports cards can be used successfully to teach mathematics, but teachers can use this as a guide to develop materials specifically suited to their classes. And now *you* are up to bat!

17

An Opinion Poll: A Percent Activity for All Students

James H. Vance

ONE of the greatest challenges facing teachers of mathematics at the elementary and middle school levels lies in providing for the needs of learners with differing abilities and interests. That individual differences among students exist at all grade levels and that the range of these differences increases from grade level to grade level is common knowledge among teachers and has been well documented by research. Jarvis (1964), for example, found that arithmetic achievement among sixth-grade students may vary by as much as seven years and that even among students with IQs of 115 or higher, the range of achievement is about five years.

Several procedures for accommodating individual differences in mathematics have been explored over the years. The homogeneous grouping of students is intended to reduce the range of ability or achievement within a class, permitting the teacher to adjust instruction to better suit the needs of more able, average, or less able students. Self-paced or individualized programs are designed to enable students to progress at their own rate of speed through a prescribed sequence of learning units.

In classrooms containing students of mixed ability in mathematics, the *multilevel performance learning activity* is an effective way of providing for individual differences. Multilevel learning activities are problems, games, or settings for computational practice that are assigned to the whole class but permit each student to work at his or her level of understanding. The task must capture the interest of the learners and be easily understood so that each student can immediately begin working in some way. For example, the activity might initially involve the use of concrete materials or require only basic counting techniques. However, the interpretation of the problem, the method of investigation or solution, and the depth to which the ideas are explored will vary with the individual student.

One type of multilevel performance learning activity is the open-ended problem that requires the students to collect, interpret, and communicate data (Liedtke and Vance 1978). An example of this kind of classroom activity is presented here with suggestions on how it might provide for learners with different abilities.

An Opinion Poll

Taking an opinion poll and charting the results is an activity that can be used as part of a sixth- or seventh-grade unit on percent. In addition to furnishing a setting in which students can apply concepts relating to percent in a real-life situation, the task involves the construction and interpretation of tables and graphs—aspects of one of the basic skill areas in mathematics.

Teacher presentation and instructions

Class members working in pairs conduct a poll to investigate the preferences of the students in their school on some topic. Each team prepares for the poll by choosing its own topic, such as favorite foods, favorite cars, favorite sports, and so forth, and providing five to nine choices for the respondents. The polling is conducted outside of class time. During the mathematics period the students prepare a poster chart to communicate the results of their survey. The chart (about 30 cm by 45 cm) includes a title, the raw data, the data in percent form, a "hundreds graph" (a ten-by-ten grid), a bar graph, and a summary statement. Students are encouraged to be imaginative in making the poster; they may use colored pens and cut pictures or words from magazines to illustrate the topic under consideration. Seventh-grade students at Torquay Elementary School in Victoria, British Columbia, conducted opinion polls that resulted in posters similar to the ones pictured in figures 17.1, 17.2, and 17.3.

Learning outcomes and individual differences

Several features of this learning activity make it appropriate for a class with a range of abilities. It provides for active involvement in an interesting and meaningful task. Most students enjoy working with a partner, and they find that cooperation is important as alternatives are discussed and decisions are made concerning the topic, the survey form, and the preparation of the chart.

To slower learners the assignment does not appear mathematically complex. Conducting the poll and tabulating the results are tasks they can readily and successfully complete. Expressing the data in percent form might be carried out with the aid of a calculator. (Many groups choose a sample of 50 or 100 people to facilitate this part of the task.) Transferring the percent data to the hundreds graph provides a check on the computation and visually reinforces the concept that the total group surveyed is 100 percent. The skills required to construct a bar graph correctly can be reviewed or retaught as

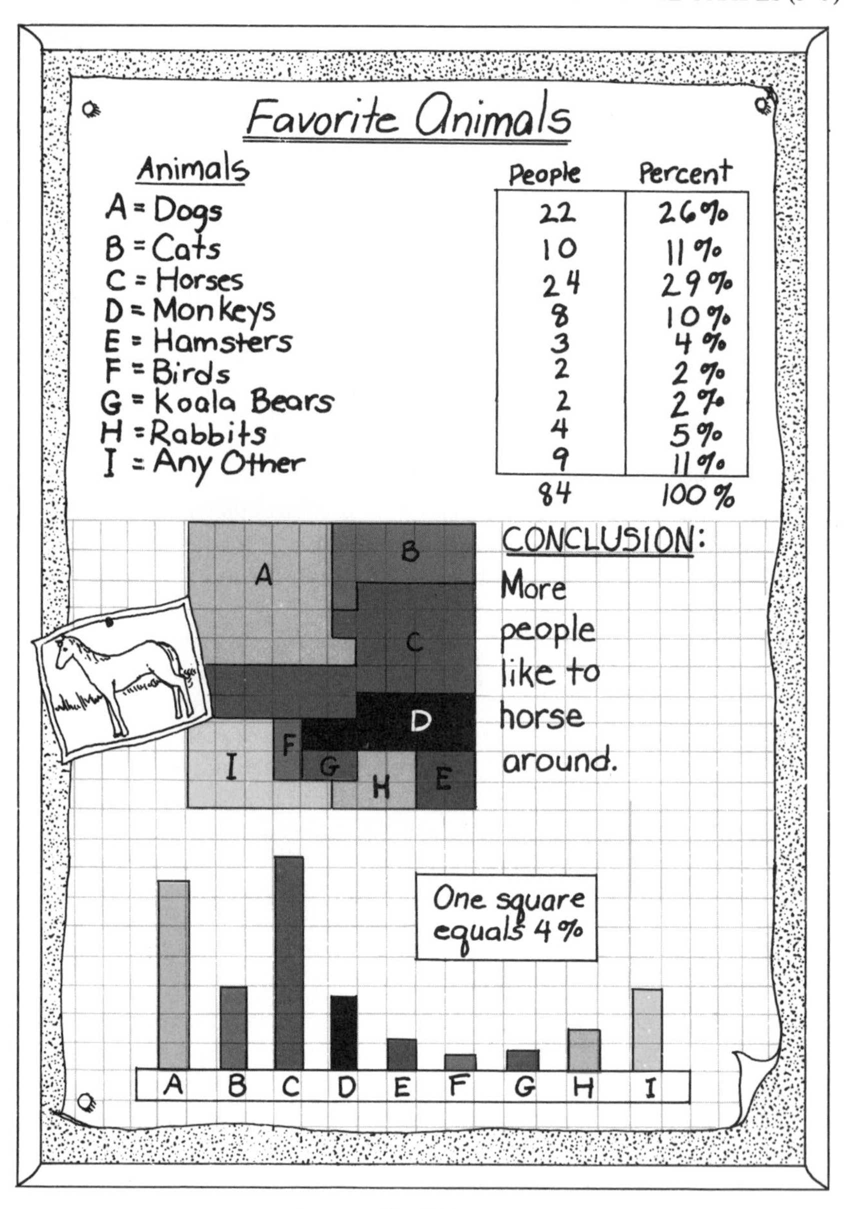

Fig. 17.1

necessary while the students work on presenting their findings in this commonly used form. Creating a title, writing a summary statement, and arranging the various items on the chart are tasks that relate to the interpretation and communication of the findings, although the students may not recognize these as mathematical activities.

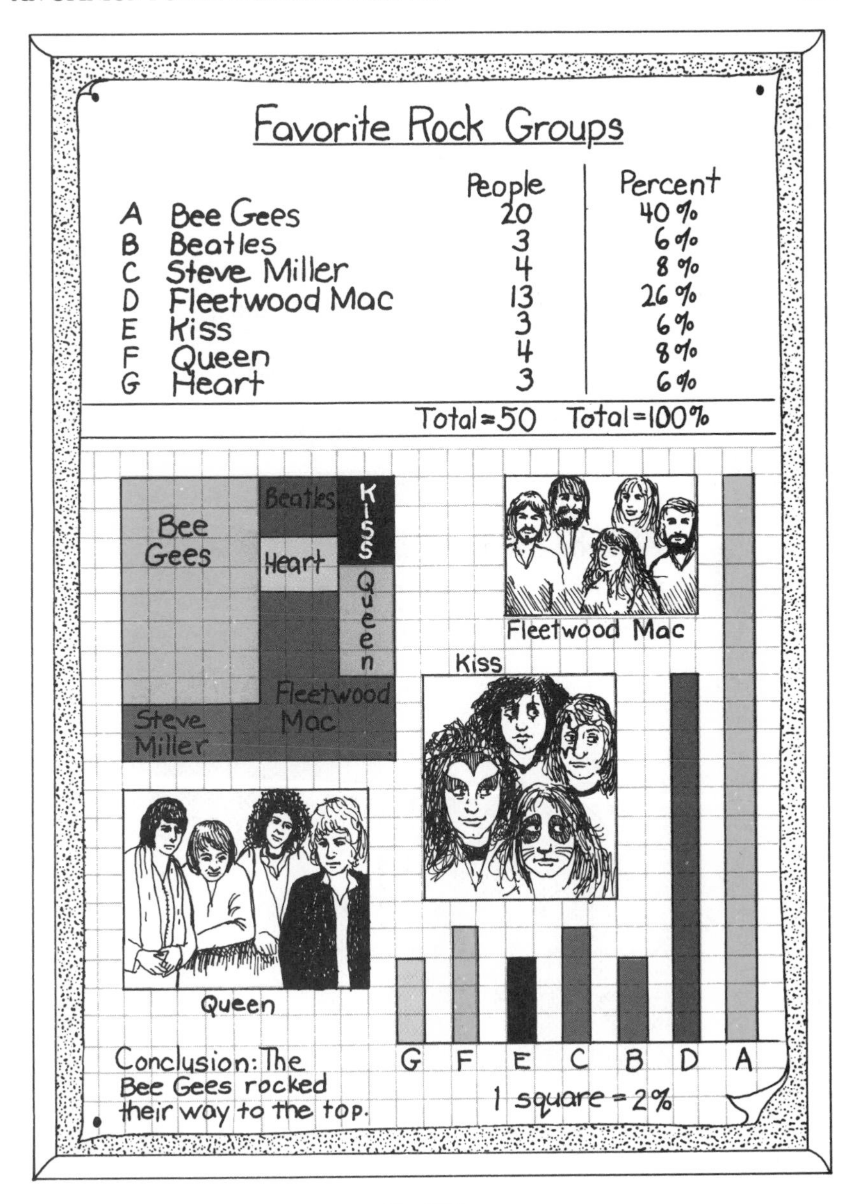

Fig. 17.2

Students with high ability in mathematics are able to work effectively at an abstract level and generally have good verbal problem-solving skills. However, these students, too, enjoy and benefit from this type of problem-solving activity. They find it a welcome change from paper-and-pencil mathematics and particularly enjoy the active and creative aspects of the task. As they carry out the assignment, they might explore ideas concerning

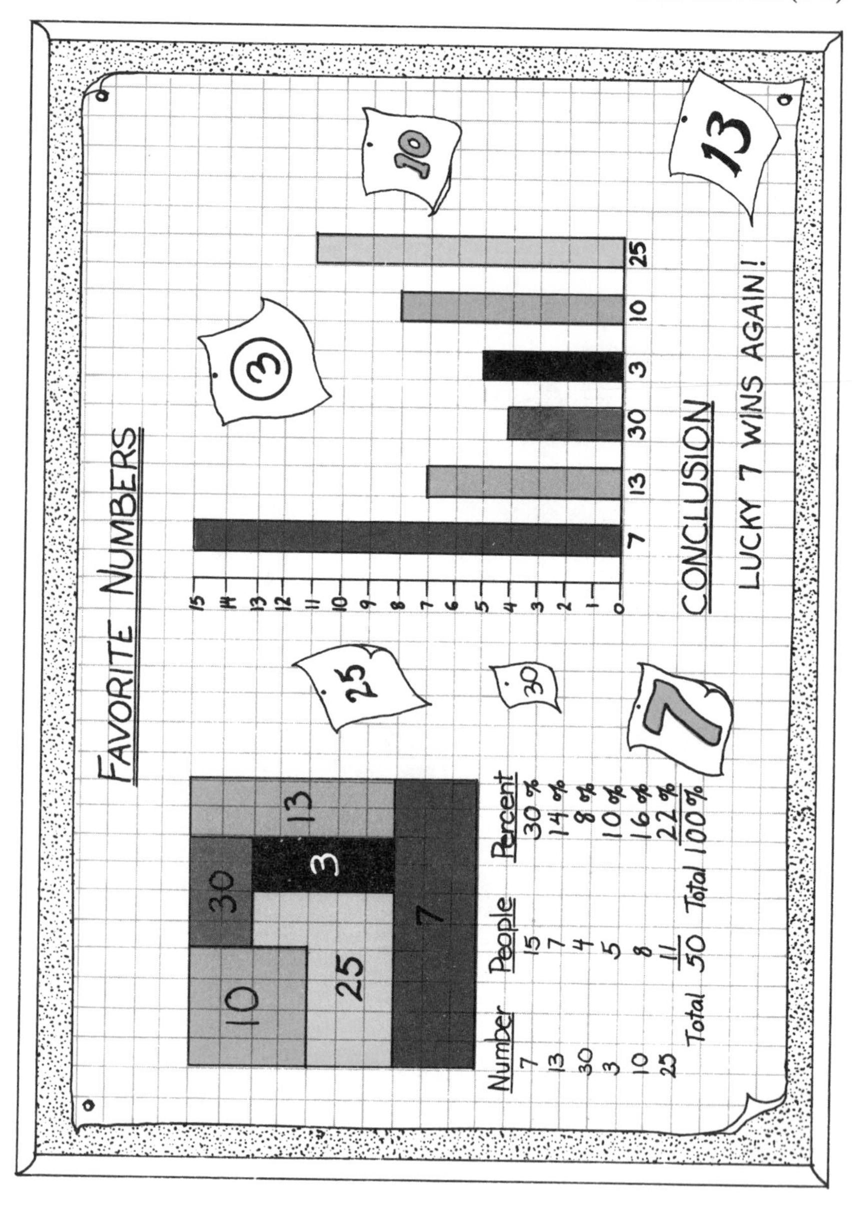

Fig. 17.3

sampling techniques and drawing conclusions from samples. The teacher can raise questions in these areas and encourage the more capable students to extend their investigation. For example, the questions "Should there be equal numbers of boys and girls polled?" and "Should all students in the sample be from the same grade?" might result in a study involving two variables, such as comparing the favorite sports of boys and girls or the favorite television programs of students at three different grade levels. Tables and graphs can be extended accordingly to display the information gathered.

As a follow-up to this project, each group of students could be asked to prepare questions that can be answered or that arise from the data presented on the chart. The level of questions might range from "Which animal was chosen least often?" and "Which three cars combined were chosen as often as the car chosen most often?" to "Do you think the results would be different for boys and girls?" and "What would you predict the results would be in another school (or with younger children, or with the same group in two years)?" The groups might be asked to present their questions orally, much as a teacher would do in a whole-class lesson. As an alternative, the questions could be written out and placed with the charts taped on the classroom walls. Students would then respond to each set of questions as they moved from chart to chart, working in a study-station format.

Summary

Activities that take into account individual differences among students in a heterogeneous, whole-class instructional setting—such as the one described here—enable slower learners to enjoy immediate success as they develop and practice basic concepts and skills. At the same time, more capable students can explore the topic in greater depth and engage in study involving higher-level mathematical processes.

Many exercises and activities in school mathematics can be adapted by the teacher to allow for multilevel performance. Two excellent ongoing sources of appropriate materials are the "Ideas" section of the *Arithmetic Teacher* (Hirsch and Meyer 1981) and the "Activities" section of the *Mathematics Teacher* (Hirsch 1980). Multilevel performance learning activities as a means of differentiating instruction in a whole-class setting can contribute in a positive way to the development of an instructional program that is *rich* for all students and *enriched* for more capable learners.

REFERENCES

Hirsch, Christian R. "Activities from 'Activities': An Annotated Bibliography." *Mathematics Teacher* 73 (January 1980): 46–50.

Hirsch, Christian, and Ruth A. Meyer. "Ideas, Ideas, Ideas from IDEAS: An Annotated Bibliography." *Arithmetic Teacher* 28 (January 1981): 52–57.

Jarvis, Oscar T. "An Analysis of Individual Differences in Arithmetic." *Arithmetic Teacher* 11 (November 1964): 471–73.

Liedtke, Werner, and James Vance. "Simulating Problem Solving and Classroom Settings." *Arithmetic Teacher* 25 (May 1978): 35–38.

18

Using Graphics to Represent Statistics

Gloria Sanok

USING graphics to represent statistics can get the student directly involved in collecting and organizing multivariate data, summarizing the information, and analyzing and interpreting the findings. Not only that, the method is fun.

Statistics are used more widely today than ever before. Extensive applications of statistics are included in the physical and biological sciences, the social sciences, sports records, business, and the advertising industry. Statistics presents numerical data as useful information. Some concept of statistics is essential not only for interpreting political (fig. 18.1), social, and economic (fig. 18.2) changes in our culture but also for understanding the way these trends influence our lives.

Children in the middle grades can benefit from an introduction to some of

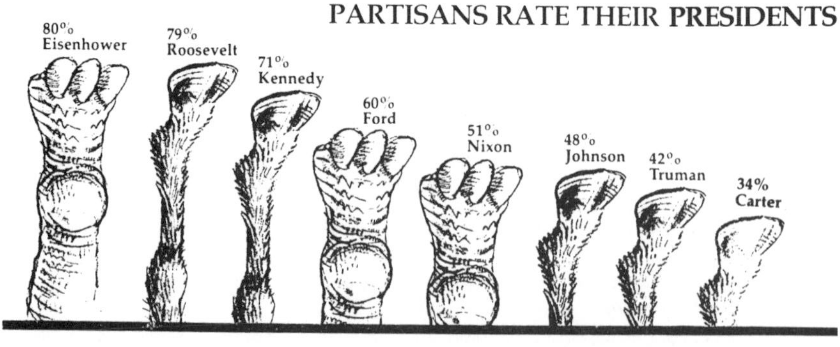

PARTISANS RATE THEIR PRESIDENTS

80% Eisenhower
79% Roosevelt
71% Kennedy
60% Ford
51% Nixon
48% Johnson
42% Truman
34% Carter

Fig. 18.1. The legs show the smallest proportion of Republicans and Democrats expressing support for a President of their own party during his time in office. Whereas Dwight Eisenhower's support in the GOP never fell below 80%, according to these data from the Gallup Organization, Jimmy Carter at his low point—summer 1979—had the approval of barely a third of Democrats. (Reprinted from *Fortune* by permission of John Huehnergarth.)

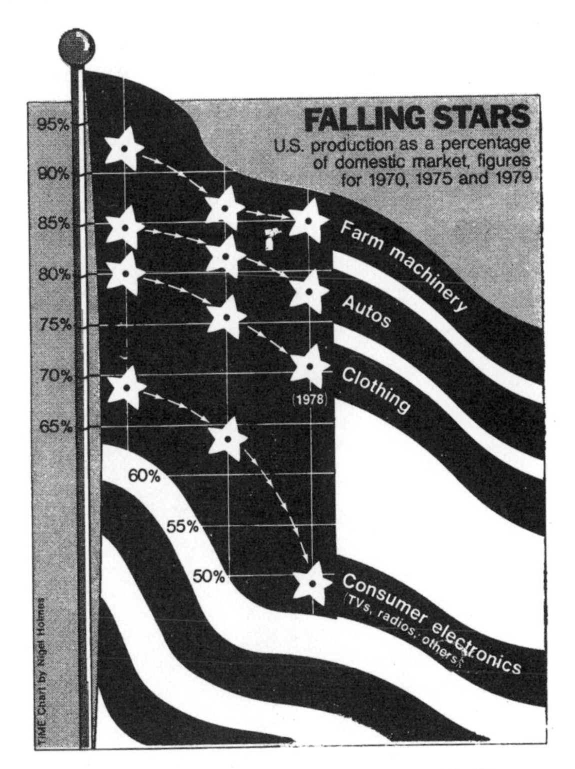

Fig. 18.2. Reprinted by permission from *Time*, The Weekly Newsmagazine; copyright *Time*, Inc., 1980.

the fundamental concepts of statistics. Learning something about how to gather and organize data should be part of each child's mathematical skills. Statistical activities in other subjects—such as recording data in science, reading and interpreting graphs in social studies, keeping and comparing charts and tables in home economics, and making scale drawings in industrial arts—will give insight into the pertinence of mathematics in many areas.

Integrating graphical representation with other subjects will provide the students with meaningful realistic examples. Children should be given opportunities to work with graphs in order to get facts and make comparisons. Many social studies and science projects offer possibilities for the pictorial or symbolic presentation of data.

The use of face representation in statistics can offer a successful approach for presenting information with many variables. This technique is effective for revealing rather involved relations that are not always visible from a single correlation based on the usual conventional two-dimensional pictographs, line graphs, bar graphs, or coordinate graphs. The faces can be used not only to aid in cluster and discrimination analyses but also to present substantial changes in a time series.

The object of this graphical method is to represent multivariate data that is subject to strong, composite relationships in such a way that an investigator can quickly understand relevant information and then apply statistical analysis. The method consists of drawing a cartoon face. A template or a stencil can be used for congruity. Variations in features—such as length of nose and curvature of mouth—correspond to pertinent information.

Each study should include a title, the legend or key (with an explanation of the detailed characteristics), the collected data, and the cartoons. Then the work can be displayed so that other members of the class will benefit.

Figures 18.3, 18.4, and 18.5 demonstrate the face-representation approach.

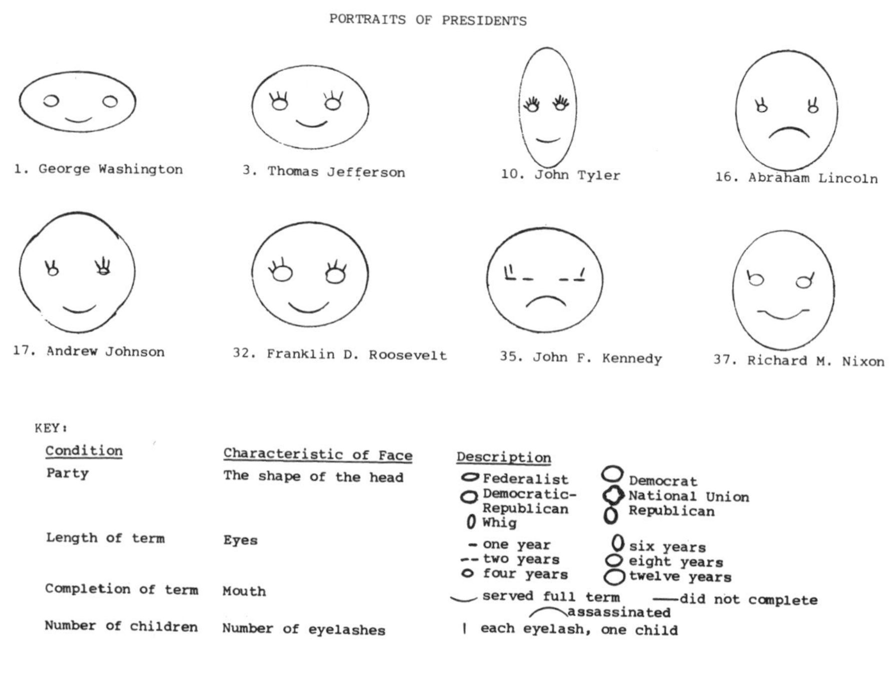

Fig. 18.3

A graph, a convenient device for organizing and presenting data, can display a lot of information at a glance. Graphs can illustrate the relationship between two or more items, as in figure 18.6.

Graphs and face representations are two means of developing mathematical skills and concepts in ordering, grouping, reading, recording, listing, counting, estimating, rounding, comparing, measuring, visualizing, identifying, recognizing, interpreting, and analyzing.

The bibliography below contains recommended reading for a more advanced use of faces in the study of statistics.

Weather Report for 16-21 April in Wayne, New Jersey

Condition	Characteristic of Face	Description
Sky	Head	The cloudier, the less round
Temperature	Mouth	The hotter, the longer
Precipitation	Eyebrows	The less the precipitation, the shorter
Wind	Nose	The higher the wind, the bigger
Wind direction	Eyes	The line points in the direction of the wind

Date	Day	Sky	Temperature		Precipitation	Wind	Wind Direction
4-16	Monday	Partly cloudy	41°F	5°C	None	2.0 rps*	Northeast
4-17	Tuesday	Cloudy	39°F	4°C	None	3.0 rps	Northeast
4-18	Wednesday	Sunny	62°F	17°C	None	2.4 rps	Northeast
4-19	Thursday	Sunny	72°F	22°C	None	2.3 rps	Northeast
4-20	Friday	Sunny	78°F	26°C	None	1.0 rps	Northeast
4-21	Saturday	Very sunny	82°F	28°C	None	0.3 rps	North

*revolutions per second

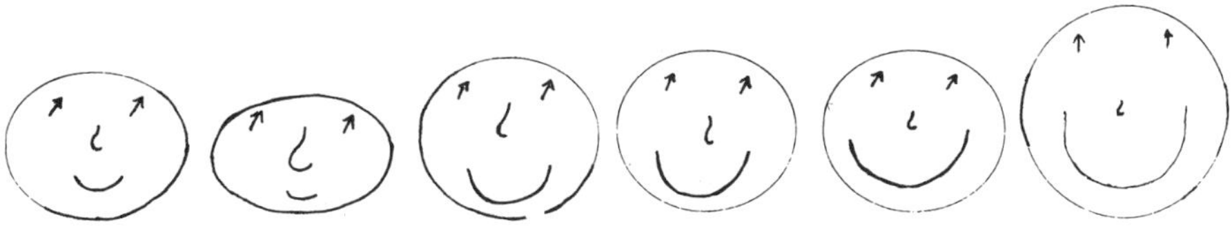

Monday 4-16 Tuesday 4-17 Wednesday 4-18 Thursday 4-19 Friday 4-20 Saturday 4-21

Jackie Dorval Grade 8
Unified Mathematics

Fig. 18.4

Weather Report for 9-15 April in Wayne, New Jersey

Condition	Characteristic of Face	Description
Temperature	Length of mouth	The hotter, the longer
Wind	Shape of head	The more wind, the rounder
Humidity	Size of nose	The more humidity, the bigger the nose
Sun	Hair	The sunnier, the more hair
Barometer	Eyebrows	The higher the reading, the bigger the eyebrows

Date	Day	Temperature	Wind	Humidity	Sun	Barometer
4-9	Monday	6°C	Little	89%	No	29.65
4-10	Tuesday	8°C	Medium	30%	Yes	30.25
4-11	Wednesday	26°C	Medium	45%	Yes	30.05
4-12	Thursday	19°C	Little	60%	No	29.95
4-13	Friday	21°C	Little	85%	No	29.87
4-14	Saturday	19°C	Much	100%	No	29.80
4-15	Sunday	17°C	Little	68%	Yes	30.21

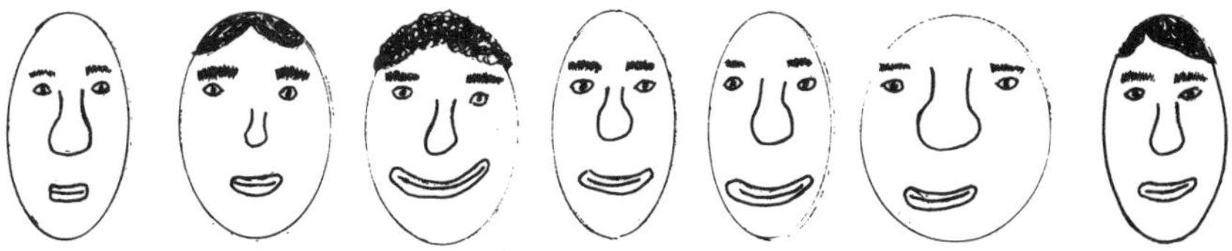

Monday 4-9 Tuesday 4-10 Wednesday 4-11 Thursday 4-12 Friday 4-13 Saturday 4-14 Sunday 4-15

Michael Lutz Grade 8
Unified Mathematics

Fig. 18.5

Student's Lunch Money

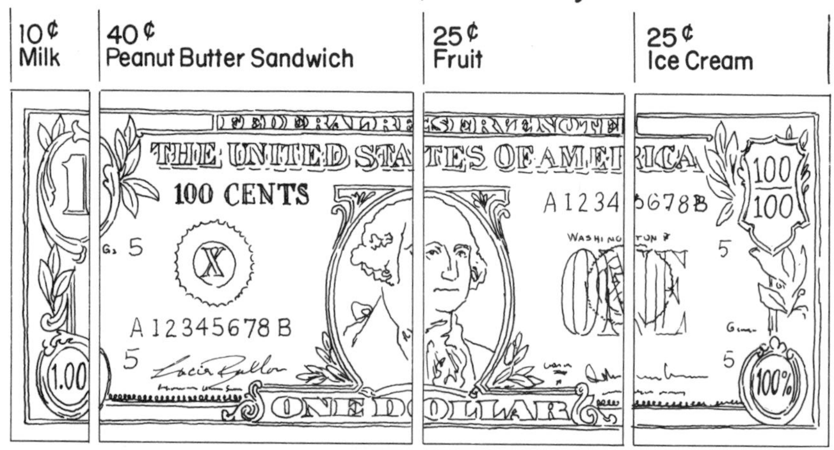

Fig. 18.6

BIBLIOGRAPHY

Chernoff, Herman. "The Use of Faces to Represent Points in k-Dimensional Space Graphically." *Journal of the American Statistical Association* 68 (June 1973): 361–68.

Jacob, Robert J. K., Howard E. Egeth, and William Bevan. "The Face as a Data Display." *Human Factors* 18 (April 1976): 189–200.

Riedwyl, Von H., and M. Schafroth. "Grafische Darstellung mehrdimensionaler Beobachtungen." *EDV in Medizin und Biologie* 7 (January 1976): 21–24.

19

Using Learning Centers in Problem Solving: Herbie and Suzie

Stephen Krulik

ONE of the most difficult problem-solving skills for students to master is determining the sufficiency of problem data. That is, is just enough information given for solving a problem? Too little information? Too much information? Are all the given facts needed? Or not needed?

The middle school marks a critical phase in the development of students' problem-solving skills. Here students make the transition from the arithmetic skills that are often emphasized in the primary grades to the conceptual development that is critical to their future mathematics education. The concept of the sufficiency of problem data must be developed here if students are to be successful problem solvers later on. At this stage, students are usually unsophisticated enough to readily accept and enjoy "Herbie and Suzie"; at the same time, they are mature enough to work independently in learning centers. Students who have used Herbie and Suzie find them fun and understandable and usually show a desire to work with them on their own throughout the school year. The choice of names—Herbie and Suzie—is a choice made by middle school students over the last few years.

What are Herbie and Suzie?

Herbie and Suzie are a pair of self-contained learning centers. They can be placed at the back of any classroom for students to use whenever they have some free time or whenever the teacher wishes to assign students to work with them. Each is made from a single sheet of posterboard (approximately 68 cm × 60 cm) that has been folded in half to form a booklet, 34 cm × 60 cm. This booklet format allows each center to stand by itself on any desk or table.

How to make Herbie and Suzie

The cover of one center (fig. 19.1) shows Herbie and the title "Needed or

Not Needed"; the cover of the other (fig. 19.2) shows Suzie and the title "Enough? Too Little? Too Much?" Inside each are the directions and a pocket that contains problem envelopes.

Fig. 19.1. "Herbie"

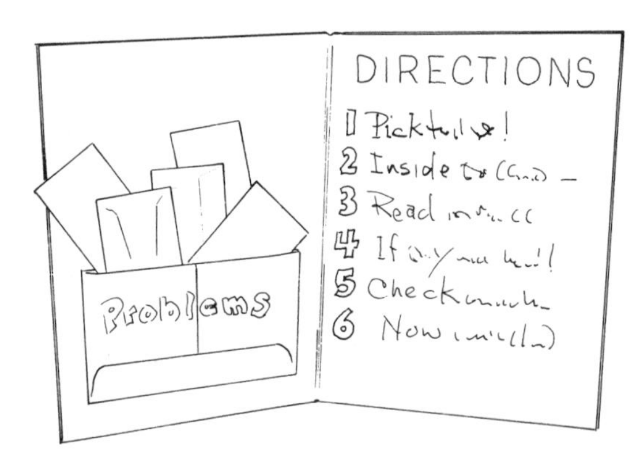

Fig. 19.2. "Suzie"

Herbie's directions are as follows:

1. Pick a problem envelope from the pocket.
2. Inside the envelope you will find a problem card and a number of fact cards. Some of the facts are needed to solve the problem, and some are not needed.
3. Read the problem. Decide which facts are needed to solve the problem and which are not needed.

4. Solve the problem.
5. Check your answer with the answer on the back of the problem card.
6. Now put all the pieces back into the envelope and put the envelope back into the pocket.

Suzie's directions are given below.

1. Pick a problem envelope from the pocket.
2. Inside the envelope you will find a problem.
3. Read the problem. Decide whether you have been given too much information, too little information, or just enough information to solve the problem.
4. If you have enough information, solve the problem. If you have too much information, pick out what you need and then solve the problem.
5. Check your answer with the answer on the back of the problem card.
6. Now put all the pieces back into the envelope and put the envelope back into the pocket.

Some typical problems

Problems for Herbie and Suzie can be collected from a variety of sources. Some might come from the students' textbook or any current textbook for grades 5–9. Information is either added (to provide "not needed" facts) or deleted (so that "too little" information is given). Other problems might come from collections of nonroutine problems (e.g., Greenes, Spungin, and Dombrowski [1977], Krulik and Rudnick [1980], NCTM [1980]). Still others can be provided by the students themselves.

Problems 1 and 2 are examples of typical problems for Herbie:

1. There are 6 rows in John's 7th-grade mathematics class. Each row contains 7 seats. How many students can be seated in John's classroom?

Fact Cards

(a) There are 6 rows in John's classroom.

(b) The class is a 7th-grade mathematics class.

(c) Each row contains 7 seats.

(d) How many students can be seated in John's classroom?

Fact (b) is not needed for solving problem 1.

2. The world's longest Monopoly game began on Monday at 9:00 A.M. and continued without stopping until 2:30 P.M. on Thursday. There were 285 players, who played 5 at a time. How many hours did the game take?

Fact Cards

(a) This was the world's longest Monopoly game.

(b) It began on Monday at 9:00 A.M.

(c) It continued without stopping.

(d) It ended at 2:30 P.M. on Thursday.

(e) There were 285 players.

(f) They played 5 at a time.

(g) How many hours did the game last?

Facts (a), (e), and (f) are not needed. In addition, to solve the problem the student must recall that there are twenty-four hours in a day.

Problems 3–5 are typical problems for Suzie.

3. Who had the highest bowling score for all three games?

Game #	Sally	Jane	Luisa
1	123	187	167
2	165	159	172
3	147	158	160

Problem 3 has just enough information for a solution.

4. The sum of the digits of a two-digit number is equal to that number with its digits reversed. Find the original number.

Problem 4 does not have enough information for finding a unique solution. As long as the units digit is zero, the tens digit can be any integer from 1 through 9.

5. Rose left her home at 8:30 A.M. and drove for 6 hours. She averaged 40 km an hour and arrived at her destination at 2:30 P.M. How far did she travel?

Problem 5 has too much information. Students should choose what is needed and then solve the problem.

Conclusion

After using the Herbie and Suzie learning centers for about six months, students find that their ability to sift out the necessary information in a given problem improves markedly. Moreover, they enjoy working with Herbie and Suzie. This was emphasized in one class when, after the teacher had taken Herbie and Suzie away in February, the entire class signed a petition for their return. (In fact, one student made stuffed figures of both Herbie and Suzie for a project in home economics class.) The teacher agreed to restore Herbie and Suzie to their previous places if the students would write a set of problem cards for each center. Writing their own problems further helped the students to develop their abilities to decide what information is needed for solving problems. Furthermore, they provided the teacher with a large collection of suitable problems to use with Herbie and Suzie for the rest of the year.

BIBLIOGRAPHY

Greenes, Carole, Rika Spungin, and Justine Dombrowski, *Problem-Mathics.* Palo Alto, Calif.: Creative Publications, 1977.

Krulik, Stephen, and Jesse Rudnick. *Problem Solving: A Handbook for Teachers.* Boston: Allyn & Bacon, 1980.

National Council of Teachers of Mathematics. *Problem Solving in School Mathematics.* 1980 Yearbook. Edited by Stephen Krulik. Reston, Va.: The Council, 1980.

20

Wanted Dead or Alive: Problem-solving Skills

Joyce Scalzitti

THE mathematics curricula of today are testimony of a serious crime in education—namely, the inability of a vast number of students to solve problems and apply mathematical concepts and algorithms to problems.

When problem solving is introduced in the classroom, it is often met with a great deal of frustration and anxiety on the part of the students. Their minds are already made up that they just cannot do it. When such a situation arises, a back-door approach to the topic becomes necessary. One such approach became apparent to me one evening while I was watching an old detective movie on television.

Isn't crime solving closely related to problem solving? And no one will argue with the fact that solving a murder is more fascinating than solving an arithmetic problem. Yet the skills employed in both are very similar. A useful approach to problem solving, then, can be constructed out of detective stories and the Parker Brothers game Clue. Once effective problem-solving skills are established, mathematical problems will no longer pose an obstacle.

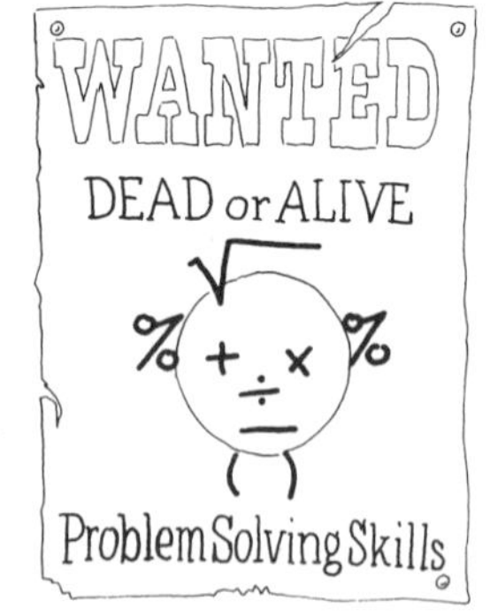

To begin, it is well worth spending at least one class period allowing the students to get the feel of being detectives by using the game Clue. This activity familiarizes the student with the thinking process a crime solver employs and lets the student feel the excitement of actually being involved in the adventure. For a class of twenty-four students, six Clue games will be needed. (It is not difficult to get six students to volunteer to supply the games. Since the game is still fairly popular, many students will have the game and will know how to play it.) The object of the game is to discover *who* committed the crime, *where* it was done, and *how* it was done. Solutions require the use of good deduction skills.

Once students have been given the taste of solving an artificial crime, the next move is to solve more realistic crimes. Crouse and Bassett (1975) discuss the use of Austen Repley's *Minute Mysteries* in developing the skills of problem solving. One is required to ask: What is given? What are you looking for? What information is needed? What information isn't needed? What pieces of information do and don't fit together logically? Take your class of detectives and put their deductive skills to work solving such mysteries as this one (Crouse and Bassett 1975, pp. 598–99):

> It was 3 A.M. as the professor and the chief of police approached the scene of the burglary. The front window was shattered. Broken glass lay all over the front porch. Inside the house Mrs. Sullivan was crying. She and her husband were dressed up and had obviously had too much to drink at the party from which they had just returned.
>
> Mr. Sullivan spoke first, "So glad you're here. My wife and I just returned from a party to find the house in shambles. My wife's jewels are missing—all heirlooms. Their cash value is great, but their sentimental value is irreplaceable."
>
> The chief of police surveyed the room. Pictures were crooked, mirrors were broken, and the drawers were emptied onto the floor. "Don't touch anything, we'll want to check for fingerprints."
>
> "That won't be necessary," interrupted the professor. "There's been no robbery."

Ask the students questions like these:

- What crime was apparently committed? (Burglary)
- What is the evidence? (Missing jewels)
- What about the scene? (House in shambles, pictures crooked, mirrors broken, drawers emptied, glass all over front porch)
- Who are the witnesses? (The Sullivans, residents of the house, who had returned from a party and had drunk too much)

Examine the information. Something there has led the professor to declare there was no robbery. How did he know? Encourage the students to go back over the details given, to picture the scene of the crime, to examine all clues carefully. Ask them if they feel they need more information than what was given. Help them to see that the key lies in the broken glass.

One or two periods should be spent working on these problems in small and large groups. Mystery stories like these can be obtained from various sources. Stress that a detective's work is methodical and follows a logical trend of thought based on evidence.

If time permits, a final activity can be employed to give further practice. Divide the class into three teams: Team 1—Police, Team 2—Suspects, Team 3—Victims. The suspects and the victims together plan and execute a mock crime, such as a burglary, without the knowledge of the police team. Certain hints or clues would be provided, and it would be the job of the police team to pick out the criminal from among the members of the suspect team. Many times one learns best by doing.

But eventually the direction must turn back to mathematical problems, and a correlation should be drawn between the two areas. In his book *How to Solve It,* Polya (1973, pp. xvi–xvii) outlines a plan against which a correlation can be drawn:

POLYA	**POLICE**
1. *Understanding the problem*	1. *Determining the crime*
What is unknown?	What crime has been committed?
What are the data?	What is the evidence?
What is the condition?	What is known about the scene?
2. *Devising a plan*	2. *Investigating the crime*
Have you seen it before?	Question people.
What key words are there?*	Examine the data.
What operations?	List the suspects.
3. *Carrying out the plan*	3. *Accusation*
Solve.	Arrest the suspect.
4. *Looking back*	4. *Trial*
Check.	Is the suspect guilty?

*Key words refer to those words from normal usage that signal appropriate mathematical operations, such as *more than, greater than, sum, plus, added to, total, and increased,* which indicate addition; *difference, how much more or less, minus, subtracted, decrease, diminished,* and *less than,* which indicate subtraction; *times, twice,* and *product,* which indicate multiplication; and *divide, quotient, partition, group,* and *share,* which indicate division.

Correlation to another plan, which is outlined in the Modern School Mathematics series (Dolciani et al. 1975), can be made as follows:

Text	**Detective**
1. Read the problem.	1. Survey the scene of the crime.
2. Decide what you are asked for.	2. What crime has been committed?
3. Look at the facts given in the problem.	3. Gather evidence.
4. Decide on the operation.	4. Investigate to find the suspect.

5. Perform the operation.	5. Arrest the suspect.
6. Check.	6. Check (trial).

In practice when one takes this approach, it is worthwhile to examine problems having both too much information and insufficient information. This unit is best taught after all operations have been studied rather than after each operation. Otherwise, a student might reason that since addition, for example, was just completed, then all the problems must involve addition. That type of reasoning can be destructive to the development of good deductive skills. No clues other than those given in the problem should be supplied.

Four examples of correlating this approach follow.

> The B Sharp music store sells portable combo organs for $599.95, mandolins for $64.95, electric guitars for $139.95, and clarinets for $169.95. Josh purchased a combo organ and a clarinet. He made a down payment of $300. What was the cost of Josh's purchase? What is the balance Josh owes? [Allen and Oldaker 1978, p. 6]

The correlated solution to this problem proceeds by asking the students what is the first thing a police detective would do. First, survey the scene of the crime (read the problem—be sure students understand words such as *down payment* and *balance*). What crime has been committed? (What am I asked to find?—cost of purchase and balance owed). Gather evidence (here encourage the student to choose only those facts that are relevant to the question to be answered; other facts should be saved if it is later decided they are also needed—organ $599.95, clarinet $169.95, down payment $300.00). Investigate to find a suspect (decide on the operation of addition followed by subtraction based on the key concepts of total and difference). Arrest the suspect (perform the operations). Trial (check to see that the total purchase is $769.90 and the balance is $469.90).

> About 100 km from Houston, Wendy stopped for gasoline. On the advice of the service attendant she decided to add a can of motor oil which cost $1.35 to her car. The cost of the gasoline purchased was $8.50. How much change did Wendy receive? [Brown and Atkinson 1978, p. 14]

Once again the dialogue proceeds as above to the point of gathering evidence. At this point the student should realize that in order to determine the change received, one must know the amount tendered. Here, a good police detective would try to uncover additional clues in order to determine the suspect. The good mathematics student realizes that to complete the problem, the amount tendered must be supplied or the problem is unanswerable. It is up to the teacher to decide whether or not the student is to assign such an amount arbitrarily.

> Cathy is going to make a dessert to serve at 7:00 P.M. If she wants to chill it at least 2 1/4 hours before serving and she needs 20 minutes to get it ready to put into the refrigerator, what is the latest time at which she must start to prepare it? [Dolciani et al. 1975, p. 73]

All detectives front and center. Survey the scene of this crime (read the problem). Decide what crime has been committed (what are you asked to find?)—*the latest preparation time.* What evidence do you have to solve this crime? (what facts are given?)—*20 minutes to prepare, 2 1/4 hours (135 minutes) to chill, dinner at 7:00 P.M.* Investigate to find the suspect (decide on the operation)—*addition and subtraction.* Arrest the suspect (perform the operation)—*20 min + 135 min = 155 min, or 2 hours 35 minutes.* Then, 7:00 − 2:35 = 4:25 P.M. Trial (check 4:25 + 0:20 = 4:45 + 2:15 = 7:00—GUILTY!). The answer is 4:25 P.M.

> The scoutmaster is organizing a camping trip for a group of boy scouts. There are 110 scouts planning to go on the trip. How many buses must the scoutmaster order if each bus seats 45 scouts?

The *crime* is to find the number of buses needed to accommodate the scouts. The *evidence* is that one bus holds 45 scouts and that there are 110 scouts involved. *Investigating* leads one to choose division as the operation: 110 divided by 45 = 2 r 20. But all is not done, because what happens to the 20 scouts left over? The scoutmaster must order 3 buses to take all the scouts. The key idea is that all 110 must go; therefore, regardless of the rules of rounding, the answer must be 3.

In all, this in-depth problem-solving unit should take about eight or ten class periods. At every opportunity, however, it should be further developed. Detective stories can be used for extra credit, challenging assignments, bulletin-board displays, and so on.

The relationship between solving detective stories and solving mathematics problems is astounding and well worth using. Turning a class of apathetic problem solvers into first-rate "detectives" is a tall order, but the fringe benefits obtained are worth the price. You may find a class of mathematical Sherlock Holmeses on your hands!

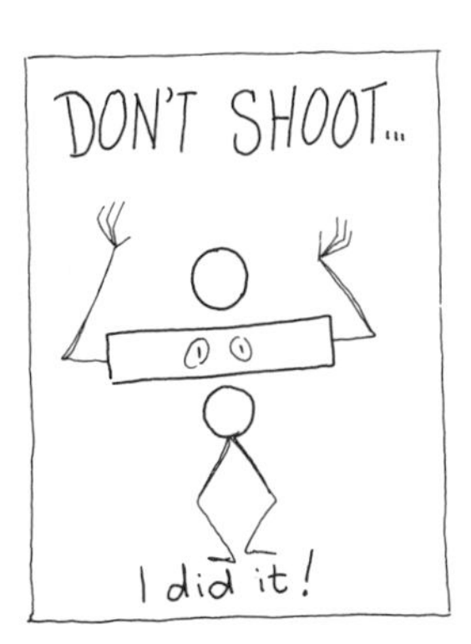

BIBLIOGRAPHY

Allen, Charles, and Linda Oldaker. *Mathematics: Problem Solving Activities.* Boston: Houghton Mifflin Co., 1978.

Brandes, Louis Grant. *The Math Wizard,* pp. 162–64. Portland, Maine: J. Weston Walch Publisher, 1962.

Brown, Richard, and Melvis Atkinson. *Mathematics: Problem Solving Activities.* Boston: Houghton Mifflin Co., 1978.

Crouse, Richard, and Denise Bassett. "Detective Stories: An Aid for Mathematics and Reading." *Mathematics Teacher* 68 (November 1975): 598–600.

Dolciani, Mary, William Wooton, Edwin Beckenbach, and William Chinn. *Modern School Mathematics—Structure and Method,* Course 2. Boston: Houghton Mifflin Co., 1975.

Polya, G. *How to Solve It.* 2d ed. Princeton, N.J.: Princeton University Press, 1973.

Sobol, Donald. *Two-Minute Mysteries.* New York: Scholastic Book Services, 1969.

21

The Flip Side of Problem Solving

Charles A. Reeves

> *When I examine myself and my methods of thought, I come to the conclusion that the gift of fantasy has meant more to me than my talent for absorbing positive knowledge.*
>
> Albert Einstein

THE record industry knows how to maximize its profits: it places a song predicted to become a hit opposite a tune with lesser expectations. This sells twice as many records as placing two hits on one record and two questionable ones on another. Rarely, but still occasionally, the tune thought to be destined for success doesn't live up to its advance billing and the flip side—the one with lesser expectations—becomes a hit. *Flip side* has consequently come to refer to something that is generally overshadowed by its neighbor. However, the flip side sometimes rises to its own level of success.

The watchword for mathematics educators in the 1980s—problem solving—has a flip side too. Some well-known rational techniques—guessing and checking, working backward, drawing diagrams, reasoning inductively, and so on—deserve, and will receive, much more attention in the mathematics classes of the present decade. But on the flip side of problem solving are those divergent, original—even autistic—mental processes that are not called on very often to resolve dilemmas but that sometimes surprise us by rising to the occasion. The most famous problem solvers throughout history—the Einsteins, the Newtons, the Galileos—have relied at times on divergent, nonlinear thought in producing their scientific breakthroughs. If we fail to expose students to this other side of problem solving in the K–12 curriculum, we may be inhibiting true problem-solving talent from reaching full fruition by failing to give the total picture.

Stimulating Creativity

Here are several activities that can expose students to the flip side of problem solving. They can also stimulate creativity in the students' approach to problem solving. You are encouraged to modify the activities to fit your personal teaching style and integrate them into your classroom over an extended period of time.

Quota

"Quota" is a way of encouraging students to think of different approaches to a single situation. This reinforces the idea that the main point of problem solving is not always to arrive at "the answer."

Put the numerals 5, 6, 7, 8, and 9 in a hat and draw one at random. On the next exercise or problem, each student or group of students must produce that many ways of solving it. For example, if "8" is drawn from the hat and the next exercise is 25 × 19, students might present the work shown in figure 21.1. If you are not sure how challenging this can be, try adding a few more ways of solving 25 × 19 to this list.

1. $\begin{array}{r} 25 \\ \underline{\times 19} \end{array}$

2. $\begin{array}{r} 19 \\ \underline{\times 25} \end{array}$

3. $\left.\begin{array}{r} 19 \\ 19 \\ 19 \\ \vdots \\ \underline{+19} \end{array}\right\} 25$

4. $\left.\begin{array}{r} 25 \\ 25 \\ 25 \\ \vdots \\ \underline{+25} \end{array}\right\} 19$

5. $[(25 \times 4) \times 19] \div 4$
   $= (100 \times 19) \div 4$
   $= 1900 \div 4$

6. $25 \times (20 - 1)$
   $= (25 \times 20) - (25 \times 1)$
   $= 500 - 25$

7. $25 \times (10 + 9)$
   $= (25 \times 10) +$ nine 25s
   $= 250 +$ four 25s $+$ four 25s $+ 25$
   $= 250 + 100 + 100 + 25$

8. $19 \times (19 + 6)$
   $= 19^2 +$ six 19s
   $= 361 + 38 + 38 + 38$
   $= 361 + 114$

Fig. 21.1

Quota works well both with exercises (as above) and with true problems. Consider these:

1. Find five easy ways to add the first 100 natural numbers.
   $(1 + 2 + 3 + 4 + \ldots + 96 + 97 + 98 + 99 + 100 = ?)$
2. Find seven ways to measure how much air your lungs hold.

The first problem typically elicits such responses as "use a calculator" and "add 1 + 99, 2 + 98, and so on, and then count the sums of 100 and anything left over." If students are pushed beyond this, they'll find a variety of ways to regroup the given addends or to calculate the sum of the digits in the ones place (450) and the tens place (also 450).

In response to the second problem, suggestions for measuring the air in someone's lungs usually start with blowing up balloons or plastic garbage bags and finding the volume, or reversing this procedure. Eventually students will reach such indirect methods as blowing on toy sailboats or fans and comparing the effect to a known volume of air or weighing oneself before and after a deep breath and dividing the difference by the density of air. Notice that practicality is not at a premium in such an experience, which is just as it should be in divergent-thought activities.

Two points are essential to this activity. The first is that the main point is not to arrive at "the answer" but rather to think of different approaches to a single situation. Second, the quota, or required number of solutions, is necessary. Merely encouraging students to "find as many as you can" allows them to quit after the first—and most obvious—two or three solutions. This strategy and the one that follows both work well when students are grouped and use the rules of brainstorming.

What if . . .?

Essential to creativity in problem solving is the willingness and ability to reshape one's thinking about apparent truths. Producing alternatives to "obviously true" basic assumptions has resulted in many of our quantum leaps in science. "What if . . .?" is designed to foster this risk-taking characteristic in students.

Initiate this activity by starting an obviously false statement with "What if. . . ." The activities that follow the "What if . . .?" premise can vary quite a bit but—as in brainstorming—should extend the idea with only positive viewpoints allowed. Such statements as those in figure 21.2 would be appropriate for the intermediate level.

1. What if our year had 100 days in it?
2. What if "487" meant "4 · 1 + 8 · 10 + 7 · 100"?
3. What if you wanted to solve "9/10 ÷ 3/2" and you didn't know to "invert and multiply"?
4. What if the word name for "15" were "onety-five"?

Fig. 21.2

The first proposal would challenge students to think of the implications of having a year with 100 days and to manipulate the number of days in a month, week, and so on, accordingly. If our numeration system were reversed, as suggested in question 2, could we still add, subtract, multiply, and divide if new algorithms were developed? What would the new algorithms be for such a system? A group of students working with question 3 would probably eventually arrive at an unusual, workable process for dividing fractions ($a/b \div c/d = a \div c/b \div d$, but a/b might need to be renamed so that $a \div c$ and $b \div d$ are both whole numbers). Example 4 frequently results in a redesigning of our word-naming system for whole numbers, one that is much more logical than the one we use. For more on this, see "A Fifth Grade's Revision of Our System of Number Names" (Pincus 1972).

The value of this activity resides in having students develop positive alternatives based on a false premise. It is difficult to cast aside previous learning in this fashion, but having to do so allows students to consider "the truth" as only one way to view the situation.

Be a magician

The intrigue of a magic show lies in its baffling our senses, convincing us that something we think is impossible may, in fact, be possible after all. One discussion of how to perform some "mathematical magic" in front of your class can be found in "Magic in the Mathematics Classroom" (Mohr and Leutzinger 1977).

Another way to expose students to some of the mysteries of mathematics involves learning some unusual algorithms for common operations and springing them on the class at just the right moment and in just the right fashion. The *doubling and halving* method of multiplying whole numbers is a good algorithm for this sort of activity (fig. 21.3).

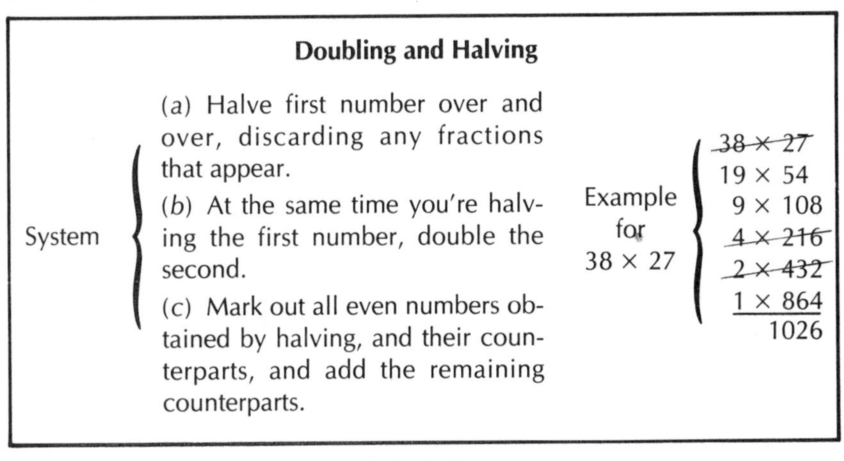

Fig. 21.3

The "magic" can also come from making use of algorithms that you understand but that students rarely see. The commutative property of multiplication allows us to compute 36×41 as shown in figure 21.4. The fact that the difference of two numbers is unchanged when the same number is added to the minuend and subtrahend justifies the subtraction process shown by the subtraction example in figure 21.4. (The main idea behind this algorithm is to add whatever number it takes to turn the subtrahend into a "nice number" to subtract.)

$$\begin{array}{r} 36 \\ \times 41 \\ \hline 246 \\ 123 \\ \hline 1476 \end{array} \qquad \left.\begin{array}{r} 5002 \\ -3684 \\ \hline \end{array}\right\} +16, \text{ becomes: } \left.\begin{array}{r} 5018 \\ -3700 \\ \hline \end{array}\right\} +300, \text{ becomes: } \begin{array}{r} 5318 \\ -4000 \\ \hline \end{array}$$

Fig. 21.4

The purpose of this technique is to demonstrate without explanation that there are some unusual ways to perform even the most familiar tasks at hand. This should bring students to accept that we need to maintain a flexible mind in judging even the most original methods we encounter for solving problems. Something we don't understand may, in fact, work just as well as something we do understand.

Some curiosities about mental processes

Certain very popular experiments related to mathematics clearly demonstrate to intermediate-level students some of the nuances of human mental processes. One of the most obvious is that when we first meet a problem we begin to impose our own restrictions.

Three quite popular problems useful for demonstrating this factor are given in figure 21.5. Try them yourself if you've never seen them before.

A farmer wants to plant 4 trees, each one the same distance from the other three. How can he do this?	Connect the 9 dots with 4 continuous line segments, without lifting your pen: • • • • • • • • •	Fill in the next letter of the alphabet, using the logic established to list the first eight. O,T,T,F,F,S,S,E ?

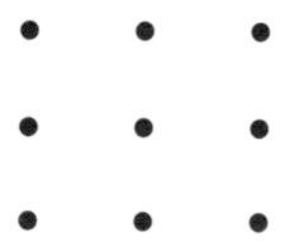

Fig. 21.5

Most likely because of our past experiences with similar problems, we immediately and subconsciously assume that—

- the four trees must be in the same plane;
- the outer dots limit our line segments;

• there is a numerical relationship structuring the sequence of letters shown.

When these limitations are removed, the problems become much easier:

• Given a hill or a hole, the trees could be planted as the vertices of a regular tetrahedron (fig. 21.6).

• One path for connecting the dots is shown in figure 21.7.

• The next letter of the alphabetical sequence is "N" for "nine."

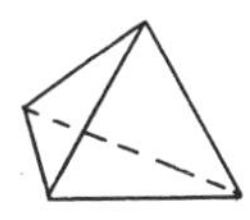

Fig. 21.6

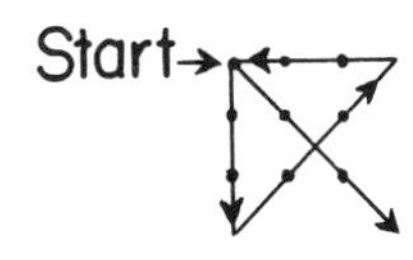

Fig. 21.7

When such a problem is to be used in a class or prior to a total-class experiment as described in the next two activities, students should not be forewarned about what's going on if the full effect is to be felt. The follow-up discussions are most effective when students have grappled with the dilemmas personally.

The famous "einstellung" experiment demonstrates to students how our minds search for patterns to apply when we are given a sequence of problems that seem to be related to each other. *Einstellung*—the German word for your mind getting in a rut—is a protective device the mind employs to free itself from thinking about a whole set of problems individually.

Have the box in figure 21.8 reproduced and let the class go through the exercises just as if they were a normal part of the assignment for that day, but allow no talking. Go over the first problem orally, and give the class ten minutes to finish the problems. The einstellung is set up, since the first few problems are most obviously solved by the same pattern ($B - A - C - C$); the mind notices this, and then tries to use this pattern on all the problems.

Imagine yourself near an unlimited water supply. Any water you don't need can be emptied on the ground. Given jugs of varying capacities, your problem is to obtain a certain amount by filling, draining, and so on. You certainly can't guess at the volume of any partially filled jugs—all you know about the volume is what it is when the jug is filled to the rim.

In the first problem below, for example, you have three jugs with the following capacities:

1 liter 27 liters 3 liters

You are asked to obtain exactly 20 liters. One way would be to fill the 27-liter jug, and then fill the 1-liter jug from it, leaving 26 liters in the large jug. Then fill the 3-liter jug two times from the larger one, leaving exactly 20 liters in the larger jug. This could be written algebraically as $B - A - C - C$.

Problem	Jug *A*	Jug *B*	Jug *C*	Obtain	Method
1	1	27	3	20	$B - A - C - C$
2	21	127	3	100	
3	14	163	25	99	
4	18	43	10	5	
5	9	42	6	21	
6	20	59	4	31	
7	23	49	3	20	
8	15	39	3	18	
9	28	76	3	25	
10	18	48	4	22	
11	14	36	8	6	

Fig. 21.8

Use the general information below to lead the follow-up discussion:

- Problem 5 is where alternative ways to $B - A - C - C$ begin to emerge. Some will notice, but most won't.
- Problem 7 is where an alternative method ($A - C$) is much shorter. Again, some won't see it.
- Problem 9 is the first case in which $B - A - C - C$ won't work. A surprising number of people can't shake their "einstellung" even at this point, and conclude that the problem can't be solved.

Mention how einstellung creeps into our everyday lives. Sitting in the same place every day in a classroom or in the lunchroom or taking the same route home from school each day are the sorts of things with which intermediate-level students can identify in their own lives. Make sure they understand that einstellung isn't a "bad" thing—we just need to be aware of its presence and influence.

Another such experiment comes from Edward de Bono (1967) and works well in a geometry unit. This experience illustrates how our minds receive knowledge from the outside world and then force it to mesh with previously acquired knowledge. The total knowledge received at each stage is structured so that it makes sense in and of itself, without reliance on any future information being revealed.

Before class, make one cardboard figure for each student that looks like the one in figure 21.9. Number and cut each one into pieces as shown. When class begins, tell the students only that they will each receive some cardboard shapes—don't tell them how many—and that their task is to arrange the shapes to form the simplest geometric figure they can and then to describe its shape to a neighbor who can't see it. Proceed as outlined in figure 21.10.

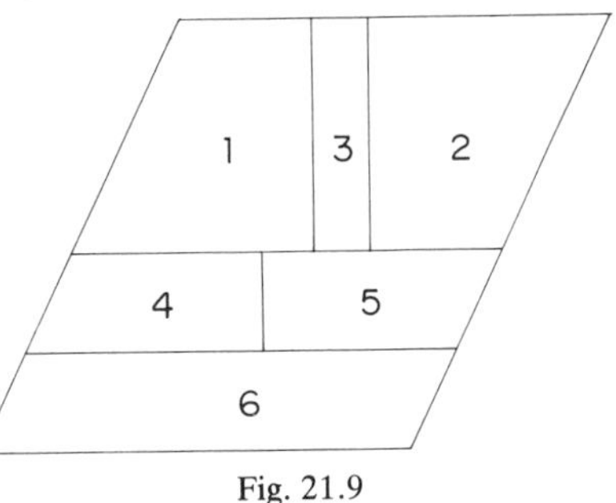

Fig. 21.9

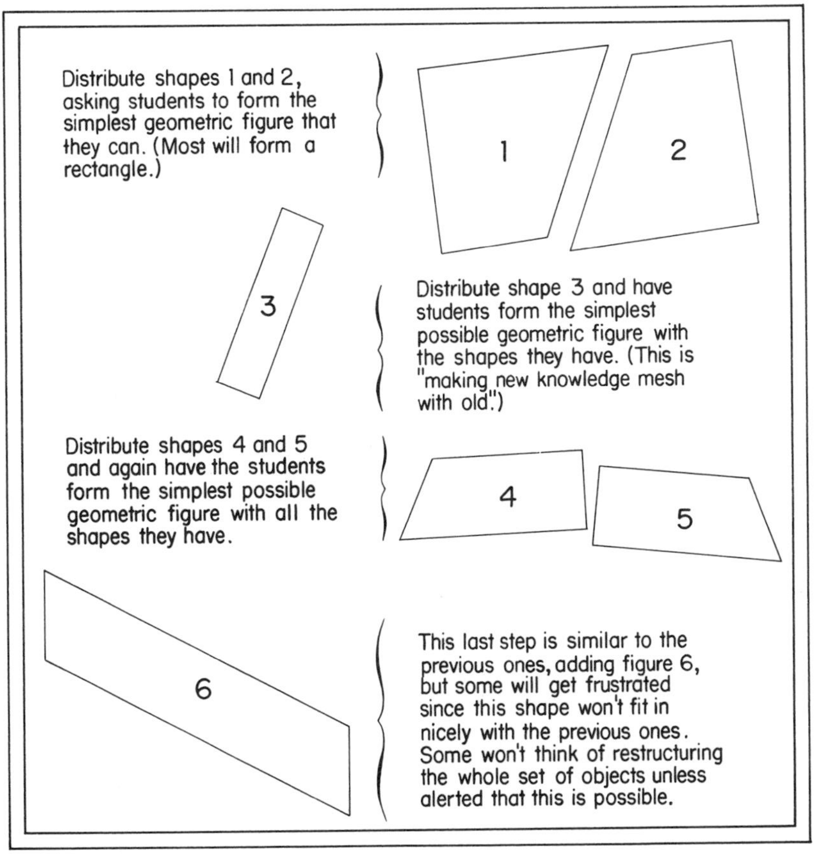

Distribute shapes 1 and 2, asking students to form the simplest geometric figure that they can. (Most will form a rectangle.)

1

2

3

Distribute shape 3 and have students form the simplest possible geometric figure with the shapes they have. (This is "making new knowledge mesh with old.")

Distribute shapes 4 and 5 and again have the students form the simplest possible geometric figure with all the shapes they have.

4

5

6

This last step is similar to the previous ones, adding figure 6, but some will get frustrated since this shape won't fit in nicely with the previous ones. Some won't think of restructuring the whole set of objects unless alerted that this is possible.

Fig. 21.10

At each stage except the last, the pieces all go together to form a rectangle, and most people will discover this. The last step—adding shape 6—produces the original parallelogram when the other pieces are rearranged. Many students who produce rectangles for the first steps of the experiment will subconsciously hold on to this shape and not restructure the figure for the last shape. This is analogous to what our minds try to do whenever we receive contradictory information about a concept that we think we already understand. The main point is to enable students to analyze and recognize their own mental processes so they can consciously alter them when the situation calls for such a drastic step.

Humor

There is a high correlation between creativity and humor. Encouraging the humor that is sometimes hidden in mathematical situations stimulates creative thought. The essence of most humor—an unusual twist to a familiar situation—is something that many students subconsciously search out in their own everyday lives. The following activity can be both fun and profitable for stimulating divergent thought in the classroom.

Such comic strips as Peanuts and B.C. and the single-frame works of Sidney Harris offer a wealth of cartoons particularly suited to mathematics and that fit in very well at the intermediate level. An interesting, stimulating activity is to post a cartoon like the one in figure 21.11 on the class bulletin board, with the punch line covered, and give students a week to write their own. Don't be surprised if some are better than the original!

Fig. 21.11. ©1979 United Feature Syndicate, Inc. Reproduced with permission.

Cartoons in particular, but other humorous ideas as well, can be used to set the stage for interesting discussions about mathematical ideas. This cartoon, for example, could be used to launch a discussion either about math anxiety or about solving problems through inductive reasoning (the class would solve simple problems involving books on a shelf and use the pattern that emerges to solve Peppermint Patti's problem).

A sample menu like the one in figure 21.12 might be used to make a point that a child will always remember simply because of the humorous answer:

Problem: How many tacos could you buy for $1?

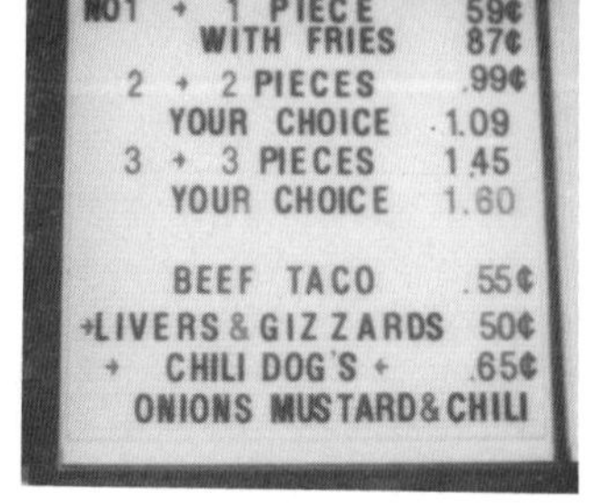

Fig. 21.12

Answer: 181 (Since each taco costs only .55¢!)

You might even try your own hand at cartooning, slipping your efforts in casually to see if the class follows your unusual twist. In having students practice drawing geometric figures, one intermediate-level teacher inserted the problem in figure 21.13—obviously an attempt at original cartooning. But it did produce the desired effect—a few chuckles, with the rest of the class wondering, "What're they laughing at?" The students then became quite interested in drawing complex pictures that were simple closed curves. The classroom humor, then, had a double payoff—stimulating divergent thought and helping students internalize a concept being studied.

Fig. 21.13. This teacher is trying to demonstrate "simple closed curves" by drawing an interesting one for the class.
Can you draw one more interesting than this one?

Puns and joke problems, such as the ones in figure 21.14, are bound to elicit moans and groans from intermediate-level students—a sure sign of success!

1. Can anyone show that half of 8 is 3?
2. You have 30 straws and 3 milk shakes. How can you put an odd number of straws in each glass, using all 30 straws?

Fig. 21.14

For these problems, erasing half of the symbol 8 gives the symbol 3; and putting ten straws in each glass yields an "odd" number in each, since you hardly ever see a milk shake with ten straws in it!

Soon some of the students will begin creating their own classroom humor, and then you have them hooked on divergent thought processes.

Fun 'n' Games II (fig. 21.15) is a comical word-matching game produced by several intermediate students on their own time. The humor comes from having to mispronounce or misinterpret the words to make them match up with the descriptive phrases.

Fun 'n' Games II

Match each statement with the correct word:

____	1. Keeps the sun off your eyes in a car.	a. zero
____	2. Telling someone to go around you.	b. divisor
____	3. Movie row right after Row Y.	c. compass
	⋮	⋮

Fig. 21.15

Besides brightening up the overall classroom atmosphere, integrating humor into the daily classroom routine enables students to appreciate and consider unusual interpretations or twists of common situations. In turn, this stimulates the ability to generate original, creative ways of viewing problems whenever they are confronted. Teachers should begin to build and use their own collections of mathematical humor and encourage students to do the same. A good book to start with is *Mathematics and Humor* (Vinik, Silvey, and Hughes 1978). Such a collection can be rapidly augmented by searching through recent mathematics textbooks that use cartoons prolifically.

Concluding Remarks

The success of any of these techniques rests to some degree on the manner in which they are presented. A low-key delivery, over a full year's time, seems to work best and enables the teacher to maintain control of the situation. In such an environment, the flip side of problem solving will flourish and become a natural part of the curriculum.

The activities described here are not meant to supplant the normal teaching of logical thought processes in mathematics. Rather, they should help internalize deductive thought by showing the "flip side." Resolving the problems of the future might require totally new perspectives, new value systems, and new theories—and only the creative problem solvers can supply these factors. Let's not neglect this aspect of problem solving in the decade of the 1980s.

REFERENCES

de Bono, Edward. *The Five-Day Course in Thinking.* New York: Basic Books, 1967.

Mohr, Dean J., and Larry P. Leutzinger. "Magic in the Mathematics Classroom." *Arithmetic Teacher* 24 (April 1977): 298–302.

Pincus, Morris. "A Fifth Grade's Revision of Our System of Number Names." *Arithmetic Teacher* 19 (March 1972): 197–99.

Vinik, Aggie, Linda Silvey, and Barnabas Hughes. *Mathematics and Humor.* Reston, Va.: National Council of Teachers of Mathematics, 1978.

22

Intuitive Equation-solving Skills

Betsey S. Whitman

PROBLEM solving continues to create anxiety for students and teachers alike. Although much has been said, written, and tried in an effort to overcome this deficiency, the National Assessment of Educational Progress reported in 1980 that between 1973 and 1978, mathematics achievement declined for all three ages tested (9-, 13-, and 17-year-olds), especially in problem-solving tasks.

The intuitive equation-solving exercises presented here are designed to help teach some of the problem-solving skills students lack. In particular, these intuitive skills enable students to read mathematical symbols and language with understanding and therefore help them overcome much of their difficulty in deciding how to solve an equation. Intuitive equation-solving skills do not take the place of the formal techniques taught in algebra; however, a general familiarity with quantitative statements comes through intuitively exploring the solutions of many equations.

Consider, then, these equations:

$$1.\ \frac{7x+12}{8}=5 \qquad 2.\ |2x+5|=4 \qquad 3.\ \frac{x^2-5}{4}=19$$

$$4.\ \sqrt{3x-1}=13 \qquad 5.\ \sqrt{5x+1}=5x+1$$

These are among the types of equations that seventh and eighth graders can learn to solve with ease. It is a matter of reading each equation as a series of questions. The "big" question is the one that the overall equation is asking. For example,

$$\frac{x^2-5}{4}=19$$

asks, "What number divided by 4 equals 19?"

Answering that question with "76," the student then sees the next question: "What number minus 5 leaves 76?" This is answered with "81," and then the student can finish the sequence of questions with, "What numbers when squared give 81?" Answering "9" and "–9," the student has solved a fractional quadratic equation on which many second-year high school algebra students founder.

Obviously, seventh and eighth graders need much practice with "one question" and "two question" equations before they can become skilled in solving the multiquestion type. Consequently, a successful procedure is to start with many examples of filling frames to answer questions. Students can be asked to fill cross-number arrays, magic squares, function machines with input-output boxes, and equations with frames (fig. 22.1). (The exercises that appear throughout this essay form a sequence of such equations, in order of difficulty. They are followed by techniques for teaching a class to solve them. Answers appear at the end of the essay.)

These first exercises are of the one-question type.

1. Fill the frames:

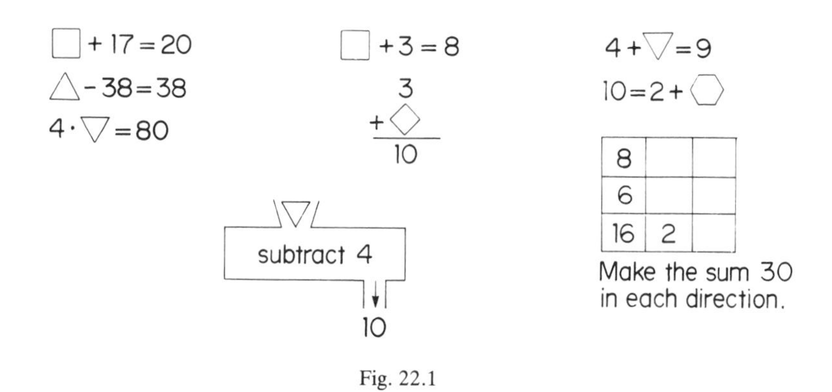

Fig. 22.1

The emphasis must be on reading the equation as a *question* to be answered with a number. Thus students must become comfortable asking, "What number plus 17 gives 20?" "What number minus 38 gives 38?" and other appropriate questions.

Then students can be introduced to expressions that contain more than one frame:

$$\square + \square = 12$$

At this point, students must learn an important rule of the game: Like shapes in the same equation require like numbers. Thus, in the equation above, "6," or "5 + 1," or some other name for the number six must be used in *both* frames.

2. Students should next try some examples such as these:

a. ⬡ + ⬡ = 16	*d.* 16 = □ · □	*g.* 5 + ◇ = 2 + ◇ + 3
b. ▽ + ▽ = 15	*e.* ○ · ○ · ○ = 8	*h.* ⬡ + 3 = ⬡
c. 1 = ◇ + ◇ + ◇	*f.* □ + □ = □	*i.* ◇ · ◇ = ◇

The last three examples need to be given special attention. The equation

5 + ◇ = 2 + ◇ + 3

asks, "Five plus what number is the same as 2 plus that number plus 3?" Students have answered the question when they point out that any number works. When looking for the answer to

⬡ + 3 = ⬡

("What number plus 3 equals that same number?"), the student must decide that *no* number works. The question

◇ · ◇ = ◇

asks, "What number times itself is equal to itself?" Here the student should observe that both 0 and 1 answer the question.

3. Next, much practice should be given with two-question equations such as the following:

a. 2 · □ + 5 = 45

b. 3 · △ − 8 = 31

c. 48 − 3 · ⬡ = 3

Much time should be spent verbalizing each instance. In these examples, it is important that students read the first one thus:

- "What number plus 5 is 45?" (40) "Two times what number is 40?" (20)
- The second equation reads, "What number minus 8 is 31?" (39) Then the next question is, "Three times what number is 39?" (13)
- The third equation is solved by asking, "Forty-eight minus what number is 3?" (45) The second question is, "Three times what number is 45?" (15)

The last equation can easily be misread. It is important that students learn to associate the product of a number and a frame as a single number in responding. The cover-up technique illustrated in figure 22.2 is helpful in overcoming the tendency to separate the number from the frame at the wrong time. Thus, if the teacher covers the

3 · ⬡

in equation 3*c*, the appropriate "big question" is seen more readily.

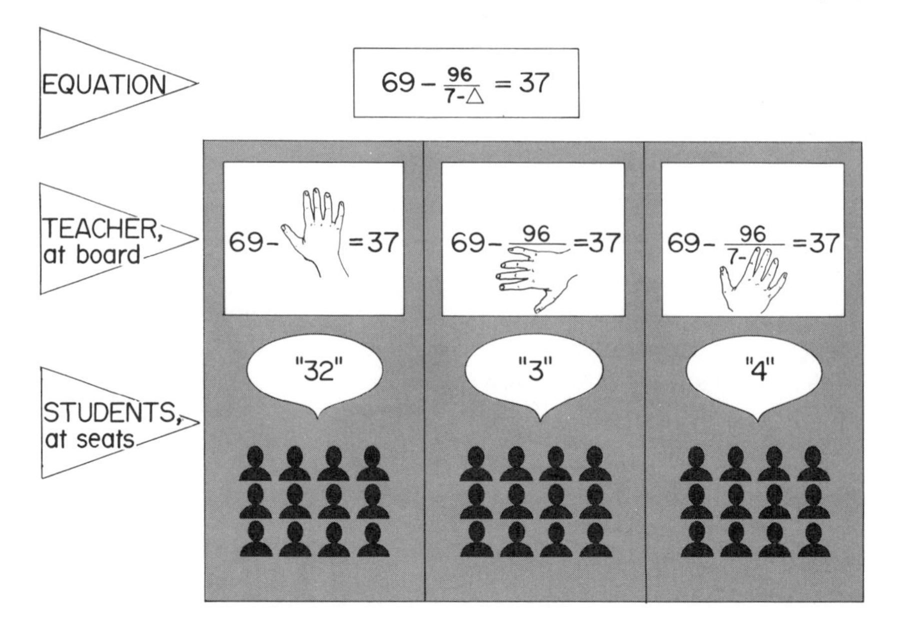

Fig. 22.2. Teacher-student interaction using cover-up technique for solving equations

Solving equations intuitively essentially eliminates the need for checking the result. Checking would consist of answering the questions in reverse order, and if students have made an arithmetic error, they are not likely to detect it in responding to the same set of questions a second time.

Several gamelike activities reinforce the filling of frames in equations. The following one has a self-checking feature:

4. Find a number that works in (*a*). Use it in (*b*). Find another number that works in the new shape. Continue.

a. $4 + \triangle = 7$

b. $9 \cdot \triangle + 2 \cdot \square = 37$ (Fill $\triangle$ by using the solution from (***a***).)

c. $4 \cdot \square = 6 \cdot \square - ⬡$

d. $3 \cdot ⬡ + 12 = 4 \cdot ⬡ + 2$

When students have learned to subtract integers, the game called The Greatest will provide a challenge. Each student is given a game sheet on which several copies of this model appear:

REJECT ⬡

$\square \cdot \bigcirc =$

$\triangle + \bigcirc =$ _______

DIFFERENCE =

One person draws four cards, one at a time, from a set that includes the numerals 1 through 10. After each draw, players write the number drawn in frames of one shape on their game sheet. Since the product and the sum have only three shapes, one number must be rejected. Each student must choose which number to reject and where to place the other numbers. After the four draws are completed, the players calculate individually their product and their sum and the difference between those two numbers. The player with the greatest difference wins. The game can be repeated several times. For more advanced students, the deck can include negative integers as well.

An easier version of the game asks for the sum as the final number. The game sheet for that version looks like this:

REJECT ⬡

□ · ◯ =

△ + ◯ = ______

SUM =

Other versions can be made up by students or the teacher.

Good techniques for eliciting considerable verbalization among students while they are learning the intuitive methods include the following:

- Use an overhead projector or blackboard to display a new list of equations each day. Different students can go to the overhead or the board and lead the class in the sequence of questions to solve a given equation. The teacher can guide and prompt a student who has difficulty.

- Have students use the cover-up technique already mentioned while they lead the class through the solution of an equation.

After eight or nine days of classroom experience with these equation-solving techniques, most students easily accept the transition to letter variables and can solve most equations in a set like this:

5. Solve:

a. $5 \cdot y + 3 = 2$ ____

b. $4 - 2 \cdot k = -(2 \cdot k - 4)$ ____

c. $4 \cdot p^2 = 100$ ____

d. $\frac{3}{5} \cdot y = 6$ ____

e. $\sqrt{4 \cdot y - 3} + 7 = 10$ ____

f. $\frac{2 \cdot n + 7}{3} = 5$ ____

g. $|x - 5| = 5$ ____

h. $\frac{\sqrt{x} + 4}{3} = 5$ ____

i. $\frac{2 + 4 \cdot |y|}{5} = 6$

When seventh graders have learned to solve these kinds of equations, they are frequently reading and understanding far more about equations than their counterparts in a second-year high school algebra class. If seventh graders learn these skills and then review them in eighth grade, they should be ready to tackle algebra in ninth grade with no qualms. In addition, they will be able to write the correct equation immediately for many verbal problems they meet in their algebra text. Many of the verbal problems in algebra texts translate easily into the kinds of equations the student who has been trained in intuitive methods has seen, and solved, before.

SOLUTIONS

1. *a.* 3 *b.* 76 *c.* 20 *d.* 5 *e.* 7 *f.* 5 *g.* 8 *h.* 14 *i.*

8	18	4
6	10	14
16	2	12

2. *a.* 8 *b.* 15/2 *c.* 1/3 *d.* 4, -4 *e.* 2 *f.* 0 *g.* any number *h.* no number *i.* 0, 1

3. *a.* 20 *b.* 13 *c.* 15

4. *a.* 3 *b.* 3, 5 *c.* 5, 5, 10 *d.* 10, 10

5. *a.* $-1/5$ *b.* any number *c.* 5, -5 *d.* 10 *e.* 3 *f.* 4 *g.* 10, 0 *h.* 121 *i.* 7, -7

III. Games, Contests, and Student Presentations

23

Using Games to Teach Fraction Concepts and Skills

George W. Bright
John G. Harvey

GAMES are frequently used in mathematics classrooms, and students perceive them as useful in learning mathematics (NAEP 1979). Given these facts, it seems important not to restrict games to drill and practice of routine skills but also to use them to teach concepts and help students acquire knowledge at higher cognitive levels (Bloom 1956).

In particular, games can be used this way to teach fraction concepts and skills. Such games should not be the total instruction, but they can provide repeated exposure to fraction ideas in a form that students readily accept. The games presented here have generally proved to increase student performance on tests typically used in classrooms.

Why use games?

Because games appeal to students, they can be effective in helping students achieve a wide range of instructional objectives. One of the most important aspects may be that while students are playing, they have control over the game-playing situation. Because of this control, they are often willing to risk making errors by trying to expand on what they know without

fear that they will fail in the eyes of the teacher or their peers. "After all," they may think, "it's only a game!"

The games presented here have three useful features. First, they can easily be modified for variety or to teach other instructional objectives. This is important, since it is often difficult to develop new games that successfully achieve a particular instructional objective. Second, each game provides feedback to students on each round of play. Because fraction concepts and skills are frequently difficult for students to master, this immediate feedback, or at least the opportunity for it, seems important. Third, and most important, is that these games are effective teaching tools. Data illustrating their effectiveness will be presented later. But whether the games are appropriate for a *particular* classroom depends on whether they provide the opportunities to reach instructional goals important for that classroom.

Fraction games

Four games, and variations of some of them, are presented: Decimal Spin, Decimal Shapes, ORTIG, and Prime Plus. The skills and concepts these games teach run the gamut of ordering common or decimal fractions, performing operations with decimal fractions, and identifying equivalents among common fractions, decimal fractions, and percents; they also have a range of cognitive levels.

Some of the games can be used for several instructional purposes. Decimal Shapes, for example, can be used for practice at the postinstructional level (after instruction designed to produce mastery) or at the co-instructional level (concurrently with such instruction).

Effectiveness of games

Decimal Spin was used at the postinstructional level with eighth graders. Students played for twenty minutes at a time, twice weekly for four weeks.

On a twenty-item test of decimal multiplication, the means increased from 9.2 to 10.9; on a forty-four-item test of decimal addition and subtraction, their means increased from 36.8 to 38.9.

Decimal Shapes was used at the co-instructional level with randomly selected seventh graders in four classes. These students played the game twenty minutes each time, eight times in a thirteen-day instructional period, along with regular instruction on decimal fractions; the remaining students in these classes received only regular instruction. On a fifty-item test on ordering decimals, the game players' mean improved from 32.1 to 47.5, and the mean of those who did not play improved from 27.9 to 45.6. There seemed to be no added effect due to game playing.

ORTIG was also used at the co-instructional level with randomly selected fifth graders in three classes. These students played the game twenty minutes each time, twice weekly for four weeks along with regular instruction on common fractions; the other students in these classes received only regular instruction. On a fifty-six-item test on ordering common fractions, the game players' mean improved from 35.6 to 50.2, and the mean of those who did not play improved from 35.1 to 46.1. There was a significant effect due to the game playing. Decimal Spin, Decimal Shapes, and ORTIG clearly seem to encourage learning.

Extending the use of good games

Experience indicates that ideas for potentially good games are sometimes difficult to develop into games whose rules are easy to understand and apply, that students enjoy playing, and that accomplish the required instructional objectives. Thus the time spent developing a new game may not be time well spent. It is, however, often easy to take an effective game and develop variations of it that seem new to students and that accomplish different instructional objectives. Variations can be developed by altering the rules of a game or by changing the equipment used to play it.

Several variations of ORTIG can be suggested. A more complete integration of the instructional objective into all phases of play results if the scoring rule is changed thus: "The score of a player on his or her turn is the largest (or smallest) common fraction in the adjoining unoccupied playing spaces." A cumulative score can be kept. If students do not yet know how to add fractions, the score on each turn can be the numerator (or denominator) of the largest (or smallest) fraction in the adjoining unoccupied playing spaces. The unoccupied playing spaces are used so that previously played pieces need not be moved; moving previously played pieces could greatly slow the play of the game and reduce the players' interest. This rule change should probably not be made until students are fairly adept at ordering fractions, since—at least at the beginning of each game—players have to order six, seven, or eight fractions to determine their score. This, too, might greatly slow the play.

Decimal Spin

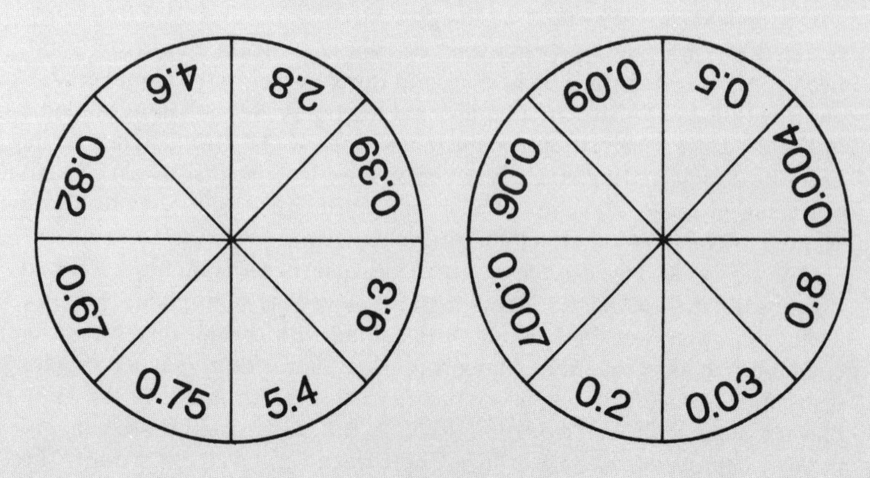

You will need

2 spinners
2 score sheets
pencils
scratch paper

Game rules

1. Take turns. When it is your turn, spin both spinners once. On your score sheet:
2. **Multiply** the two numbers on which the pointers stop.
3. **Add** the same two numbers.
4. **Add** the digit in the tenths place of the product (rule 2) to the digit in the tenths place of the sum (rule 3). This is your score for the round.

   Example: Suppose you spin 5.4 and 0.007. The product is 0.0378, and the sum is 5.407. Your score is 4 because 0 + 4 = 4.
5. Keep a running score.
6. Play 8 rounds. The winner is the player whose **grand score** is largest. In case of a tie, play one more round to break the tie.

Decimal Shapes

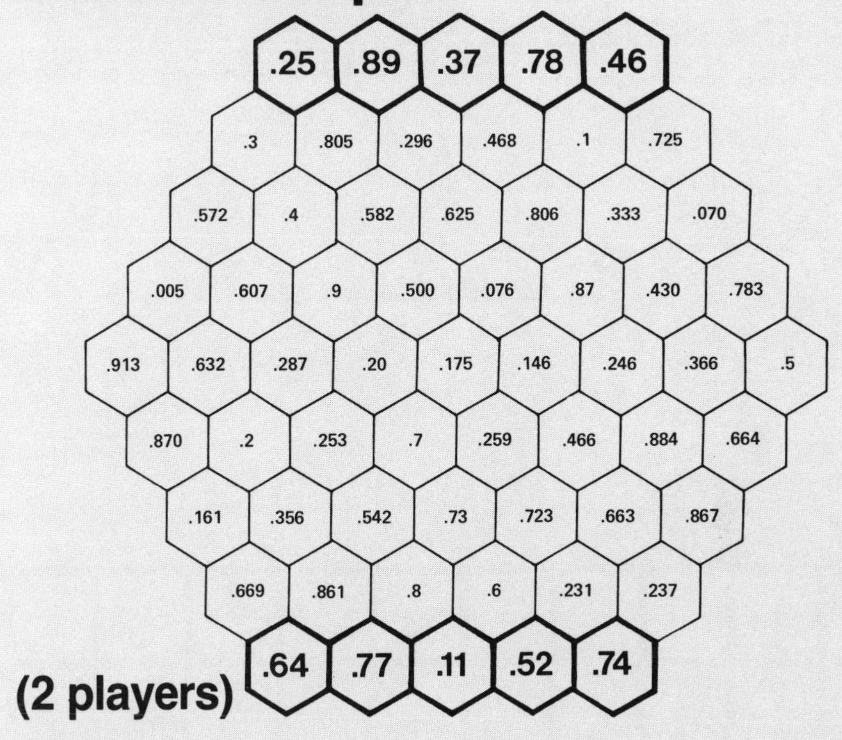

(2 players)

You will need

5 markers of a single color for each player
1 chip marked **L** on one side and **S** on the other

Game rules

1. Choose your side of the board. Place your markers on the starting positions (enclosed with darker lines).
2. Take turns.
3. When it is your turn, flip the chip once.
4. If the chip comes up **L**, move one of your markers to an open space having a number larger than the number the marker is on. If the chip comes up **S**, move one of your markers to an open space having a number smaller than the number the marker is on.
5. Legal moves are only as shown.
6. If one of your markers can move to a space occupied by your opponent's mrker, her or his marker is moved back to a starting position. Only one marker may be on a space at one time.
7. You must move one of your markers, no matter what the direction, if you are able to do so. If you cannot move, you lose the turn.
8. The winner is the first player to get all of her or his markers to the starting positions on the other side of the board.

ORTIG
(2 to 4 players)

You will need

2 numbered dice
1 die labeled A, B, C, D, E, F
1 die labeled L, L, L, S, S, S
counting chips

$\frac{2}{3}$	$\frac{1}{2}$	$\frac{8}{9}$	$\frac{4}{6}$	$\frac{4}{12}$	$\frac{3}{6}$	$\frac{2}{3}$	$\frac{3}{10}$	$\frac{1}{5}$
$\frac{2}{4}$	$\frac{7}{12}$	$\frac{4}{9}$	$\frac{1}{3}$	$\frac{3}{4}$	$\frac{3}{7}$	$\frac{8}{10}$	$\frac{6}{12}$	$\frac{1}{3}$
$\frac{6}{9}$	$\frac{3}{4}$	$\frac{1}{8}$	$\frac{1}{3}$	$\frac{4}{5}$	$\frac{5}{8}$	$\frac{8}{10}$	$\frac{4}{9}$	$\frac{8}{12}$
$\frac{3}{8}$	$\frac{2}{7}$	$\frac{3}{6}$	$\frac{3}{7}$	$\frac{8}{12}$	$\frac{4}{8}$	$\frac{2}{7}$	$\frac{2}{5}$	$\frac{3}{8}$
$\frac{6}{12}$	$\frac{3}{10}$	$\frac{6}{12}$	$\frac{6}{9}$	$\frac{3}{7}$	$\frac{8}{9}$	$\frac{4}{6}$	$\frac{1}{4}$	$\frac{4}{12}$
$\frac{2}{3}$	$\frac{1}{5}$	$\frac{5}{6}$	$\frac{1}{4}$	$\frac{3}{8}$	$\frac{5}{8}$	$\frac{5}{6}$	$\frac{6}{8}$	$\frac{5}{7}$
$\frac{2}{5}$	$\frac{2}{9}$	$\frac{3}{5}$	$\frac{2}{7}$	$\frac{2}{4}$	$\frac{6}{7}$	$\frac{2}{12}$	$\frac{1}{2}$	$\frac{1}{8}$
$\frac{4}{6}$	$\frac{2}{6}$	$\frac{1}{7}$	$\frac{2}{6}$	$\frac{5}{7}$	$\frac{6}{8}$	$\frac{1}{8}$	$\frac{6}{7}$	$\frac{4}{5}$
$\frac{7}{12}$	$\frac{7}{10}$	$\frac{1}{7}$	$\frac{7}{10}$	$\frac{2}{12}$	$\frac{2}{9}$	$\frac{2}{12}$	$\frac{3}{5}$	$\frac{4}{8}$

	1	2	3	4	5	6
A	$\frac{1}{3}$	$\frac{4}{5}$	$\frac{3}{6}$	$\frac{4}{9}$	$\frac{4}{8}$	$\frac{2}{5}$
B	$\frac{3}{4}$	$\frac{6}{12}$	$\frac{4}{12}$	$\frac{8}{12}$	$\frac{7}{10}$	$\frac{1}{7}$
C	$\frac{6}{9}$	$\frac{2}{6}$	$\frac{3}{7}$	$\frac{1}{4}$	$\frac{6}{7}$	$\frac{6}{8}$
D	$\frac{5}{7}$	$\frac{1}{5}$	$\frac{1}{8}$	$\frac{3}{8}$	$\frac{8}{9}$	$\frac{5}{8}$
E	$\frac{1}{2}$	$\frac{2}{3}$	$\frac{2}{4}$	$\frac{3}{10}$	$\frac{2}{12}$	$\frac{7}{12}$
F	$\frac{4}{6}$	$\frac{8}{10}$	$\frac{2}{9}$	$\frac{3}{5}$	$\frac{5}{6}$	$\frac{2}{7}$

Game rules

Players take turns. When it is your turn, toss all four dice. The die labeled A, B, C, D, E, F and the two numbered dice will indicate your two spaces on the grid. (For instance, if you roll A, 3, and 5, your two spaces would be A–3 and A–5.) If the two numbered dice show the same numbers, toss them again until they show different numbers. Look at the two fractions in your two spaces and decide which is the larger and which is the smaller. If **L** is showing, find the larger fraction on the game board and cover it with a chip. If **S** is showing, find the smaller fraction on the game board and cover it with a chip. If the fractions are equal, cover either of the fractions on the game board. If you cannot find an uncovered playing space to cover, you lose your turn.

Scoring

Count the number of covered playing spaces that touch a side or corner of the space you cover. Count the number of uncovered playing spaces that touch a side or corner of the space you cover. Your score for the round is the larger of these two numbers. The winner is the player with the most points at the end of ten rounds.

Challenges

Opponents may challenge a player at any time before the next player tosses the dice. If a challenge is successful, the player challenged loses his or her turn. If a chip has been placed on the game board, it must be removed.

prime plus

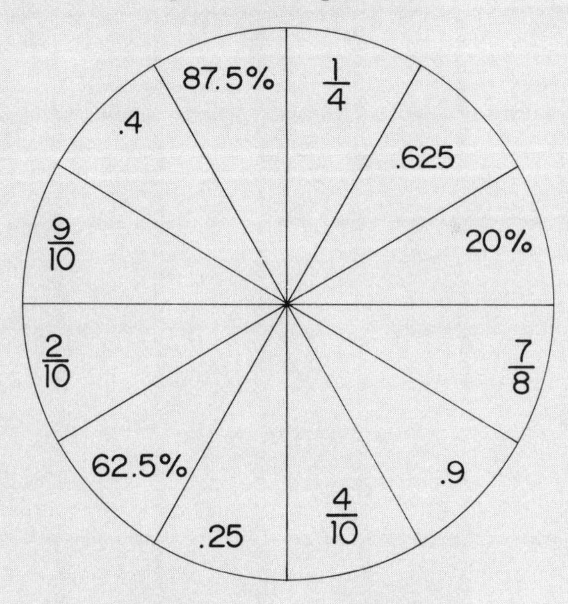

	1	2	3	4	5	6
6	$\frac{18}{20}$	$\frac{25}{40}$	.20	$\frac{5}{25}$	.900	$\frac{4}{20}$
5	$\frac{14}{16}$	$\frac{2}{8}$	$\frac{15}{24}$	25%	$\frac{42}{56}$	90%
4	.90	$\frac{3}{12}$	$\frac{27}{30}$	$\frac{2}{5}$	.875	$\frac{8}{20}$
3	$\frac{30}{48}$	$\frac{10}{16}$	$\frac{5}{8}$	.200	$\frac{28}{32}$	$\frac{5}{20}$
2	$\frac{3}{15}$	$\frac{35}{40}$	40%	$\frac{6}{15}$	.40	$\frac{21}{24}$
1	.400	$\frac{20}{32}$	$\frac{1}{5}$	$\frac{4}{16}$	$\frac{36}{40}$	.250
	1	2	3	4	5	6

You will need

spinner
2 numbered dice
paper and pencil

Game rules

1. At the beginning of each round, each player spins the spinner once to get his or her spin number for that round.
2. When all players have spin numbers, each player takes a turn rolling the dice. After each roll of the dice all the players look at the fractions or percents in the boxes determined by the dice. Each player decides if the fractions or percents in the two boxes are equivalent to his or her spin number for that round. For example, if a player rolls a 1 and a 6, every player looks at the fractions or percents in box 6,1 and box 1,6 and decides whether either fraction or percent is equivalent to his or her spin number.
3. After deciding whether either fraction or percent is equivalent to his or her spin number, each player writes down his or her score for that roll.
4. A new round begins when each player has rolled the dice once.

Scoring

Variation A

1. If you rolled the dice and one or both of the fractions or percents are equivalent to your spin number, score 2 points.
2. If someone else rolled the dice and one or both of the fractions or percents are equivalent to your spin number, score 1 point.
3. The winner is the first player to score at least 7 points.

Variation B

1. If you rolled the dice and both of the fractions or percents are equivalent to your spin number, score 4 points. Score 2 points if only one of the fractions or percents is equivalent to your spin number.
2. If someone else rolled the dice and one or both of the fractions or percents are equivalent to your spin number, score 1 point.
3. The winner is the first player to score at least 13 points.

Three changes in equipment produce other variations of ORTIG:

1. The addition of another lettered die permits common fractions on different rows of the grid to be randomly selected and greatly increases the possible number of comparisons.

2. Manipulative or pictorial materials, such as a fraction table (see fig. 23.1), which permit more direct comparison of the pairs of fractions, can be used when students are just beginning to order fractions.

3. Spinners can be used instead of dice.

ORTIG can easily be varied to change the instructional objectives as well. If the objective is to order *decimal* fractions, change the common fractions to two-, three-, or four-place decimals. To change the game so that students both order and add (or subtract) common fractions, modify the scoring rule so that the score on a given turn is the sum (or difference) of the largest and smallest fractions in the adjoining unoccupied playing spaces. (The score is then zero when all or all but one of the adjoining spaces are occupied.)

Just as the common fractions on an ORTIG board can be replaced by decimal fractions, the decimal fractions on any Decimal Shapes board can be replaced by common fractions. Experience in varying games indicates that this variation of Decimal Shapes is one that would not change student interest or ability to understand and apply the game rules. The shape of the playing board and the fractions appearing on these boards can also be changed to make possible a greater variety of comparisons than if a single board were used repeatedly. (See variation of Decimal Shapes in figure 23.2.)

Prime Plus requires students to choose from among the algorithms for converting common fractions, decimal fractions, and percents. The game is more sophisticated than Decimal Spin, Decimal Shapes, and ORTIG and can be used at the co-instructional and postinstructional levels. Prime Plus can easily be varied to deal with repeating rational representations by replacing the numbers on the game boards with common fractions that are equivalent to repeating (rather than terminating) decimal fractions, the equivalent repeating decimal fractions, and percents. Other versions of Prime Plus akin to ORTIG and other members of the "TIG" family (Broadbent 1972) can be generated to involve equivalence and operations on any combination of common fractions, decimal fractions, and percents.

These games have generally proved to increase student performance on tests typically used in classrooms, and limited modifications should be equally effective. Such variations can be developed with minimal investment of time by the teacher and can be tailored to fit specific learning needs. More extensive modifications could be attempted when time and facilities permit.

Fraction Table

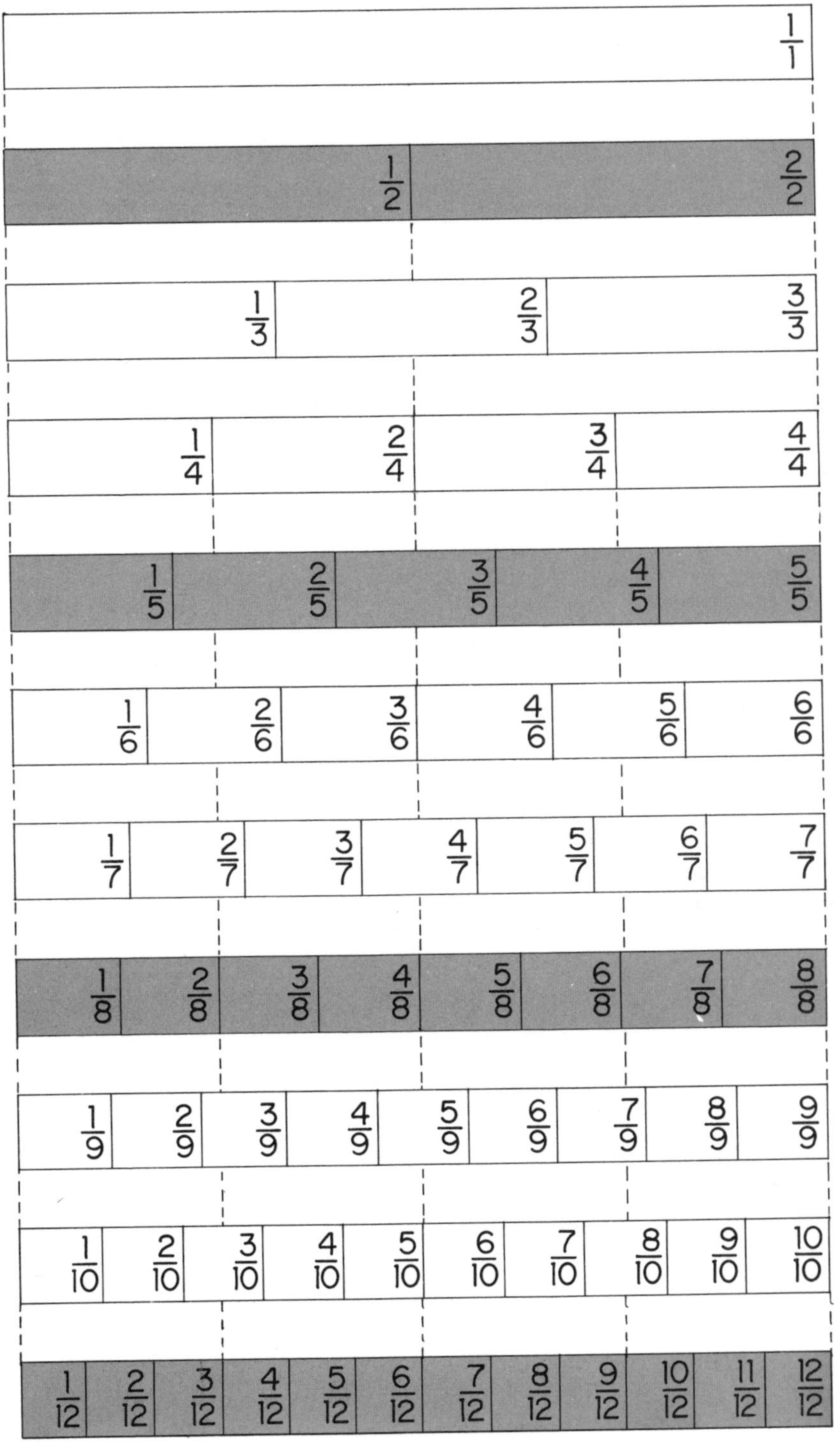

$\frac{1}{1}$

$\frac{1}{2}$ $\frac{2}{2}$

$\frac{1}{3}$ $\frac{2}{3}$ $\frac{3}{3}$

$\frac{1}{4}$ $\frac{2}{4}$ $\frac{3}{4}$ $\frac{4}{4}$

$\frac{1}{5}$ $\frac{2}{5}$ $\frac{3}{5}$ $\frac{4}{5}$ $\frac{5}{5}$

$\frac{1}{6}$ $\frac{2}{6}$ $\frac{3}{6}$ $\frac{4}{6}$ $\frac{5}{6}$ $\frac{6}{6}$

$\frac{1}{7}$ $\frac{2}{7}$ $\frac{3}{7}$ $\frac{4}{7}$ $\frac{5}{7}$ $\frac{6}{7}$ $\frac{7}{7}$

$\frac{1}{8}$ $\frac{2}{8}$ $\frac{3}{8}$ $\frac{4}{8}$ $\frac{5}{8}$ $\frac{6}{8}$ $\frac{7}{8}$ $\frac{8}{8}$

$\frac{1}{9}$ $\frac{2}{9}$ $\frac{3}{9}$ $\frac{4}{9}$ $\frac{5}{9}$ $\frac{6}{9}$ $\frac{7}{9}$ $\frac{8}{9}$ $\frac{9}{9}$

$\frac{1}{10}$ $\frac{2}{10}$ $\frac{3}{10}$ $\frac{4}{10}$ $\frac{5}{10}$ $\frac{6}{10}$ $\frac{7}{10}$ $\frac{8}{10}$ $\frac{9}{10}$ $\frac{10}{10}$

$\frac{1}{12}$ $\frac{2}{12}$ $\frac{3}{12}$ $\frac{4}{12}$ $\frac{5}{12}$ $\frac{6}{12}$ $\frac{7}{12}$ $\frac{8}{12}$ $\frac{9}{12}$ $\frac{10}{12}$ $\frac{11}{12}$ $\frac{12}{12}$

Fig. 23.1

Decimal Shapes
(2 players)

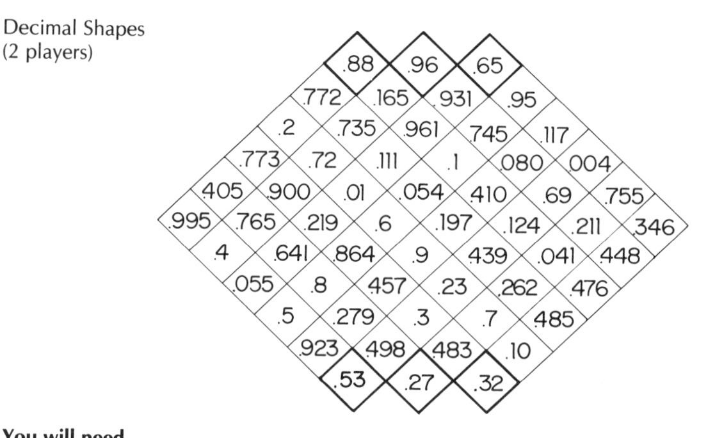

You will need

3 markers of a single color for each player
1 chip marked **L** on one side and **S** on the other

Game rules

1. Choose your side of the board. Place your markers on the starting positions (enclosed with darker lines).
2. Take turns.
3. When it is your turn, flip the chip once.
4. If the chip comes up **L**, move one of your markers to an open space having a number larger than the number the marker is on. If the chip comes up **S**, move one of your markers to an open space having a number smaller than the number the marker is on.
5. Legal moves are only as shown.
6. If one of your markers can move to a space occupied by your opponent's marker, her or his marker is moved back to a starting position. Only one marker may be on a space at one time.
7. You must move one of your markers, no matter what the direction, if you are able to do so. If you cannot move, you lose the turn.
8. The winner is the first player to get all of her or his markers to the starting positions on the other side of the board.

Fig. 23.2. Variation of Decimal Shapes

REFERENCES

Bloom, Benjamin S., ed. *Taxonomy of Educational Objectives.* New York: David McKay Co., 1956.

Broadbent, Frank W. "'Contig': A Game to Practice and Sharpen Skills and Facts in the Four Fundamental Operations." *Arithmetic Teacher* 19 (May 1972): 388–90.

National Assessment of Educational Progress (NAEP). *The Second Assessment of Mathematics, 1977–78, Released Exercise Set.* Denver: The Assessment, 1979.

24

SCINO: A Game on Scientific Notation

William J. Collins

SCIENTIFIC notation is a topic of middle school mathematics that provides ready application for such topics as place value, exponents, and operations with common and decimal fractions. However, the skill of translating between scientific and standard notation is one that itself needs practice. SCINO is a game that allows students to practice these translation skills in a competitive setting. To be successful at the game, the student must (1) translate a number from scientific notation to standard form, and (2) recognize the interval within which that number falls.

Playing the Game

SCINO can be played in large or small groups. Each student is given a copy of the SCINO board (fig. 24.1) and some chips or markers numbered from 1 to 20. These markers can be made out of oaktag or any similar material available. Three dice are needed to play. A SCINO worksheet (made up of multiple copies of fig. 24.2), although not necessary, can be a useful aid for the students, particularly as the game is being introduced.

The three dice are rolled by the teacher or by a selected student (perhaps the winner of the previous game). The numbers are called out to the students and listed on the chalkboard for future reference. The students use the three numbers called, in any order, to construct a number in scientific notation. The three numbers are used as the ones place, the tenths place, and the exponent of 10.

For example, if on the first roll the dice turn up 1, 3, and 5 (as in fig. 24.3), one student might choose to use that combination as 3.5×10^1; another might choose 1.3×10^5. The first choice, which translates to 35 in standard notation, allows the student to place his or her marker numbered "1" in the

SCINO

A between 1 and 50	**B** between 51 and 100	**C** between 101 and 500	**D** between 501 and 1000
E between 1001 and 5000	**F** between 5001 and 10 000	**G** between 10 001 and 50 000	**H** between 50 001 and 100 000
J between 100 001 and 500 000	**K** between 500 001 and 1 000 000	**L** between 1 000 001 and 5 000 000	**M** between 5 000 001 and 10 000 000

Fig. 24.1

Roll Number ______________	
□ . △ × 10⬡	Value
□ . △ × 10⬡	
□ . △ × 10⬡	
□ . △ × 10⬡	
□ . △ × 10⬡	
□ . △ × 10⬡	
□ . △ × 10⬡	
Box Used (Letter) __________	

Fig. 24.2

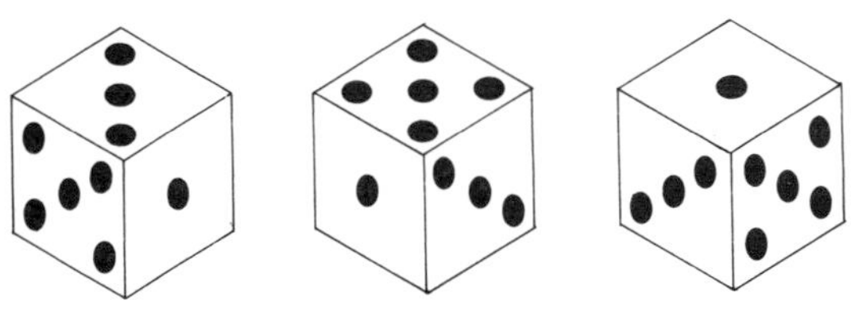

Fig. 24.3

upper left-hand box marked A on the game board. The second choice, which translates to 130 000, would allow the marker to be placed in the lower left-hand box, J. With the same combination, three other boxes might have been chosen. Since each student has one numbered marker for each roll of the dice, each student must decide which of the eligible cells will be marked for a given roll. Once the cell is chosen, the marker may not be moved. When a cell or box is marked with a numbered chip, the appropriate information should be recorded on the worksheet in figure 24.2. The game continues until one student calls "SCINO," signifying that she or he has three markers in a row.

Students should be cautioned to leave their markers on their boards until the win has been verified. To prove a win, the student must explain the three in a row, referring to three of the sets of numbers on display and how each of them was used.

Conclusion

Drill and practice in a competitive format has been proved to be effective. SCINO gives students an opportunity to practice the skill of translating from scientific notation in a gamelike setting. Exciting drill, if mathematically sound, is effective drill.

25

Variations of a Game as a Strategy for Teaching Skills

Bat-Sheva Ilani
Naomi Taizi
Maxim Bruckheimer

THE mathematics group of the Science Teaching Department of the Weizmann Institute has developed a curriculum for the junior high school. As part of the curriculum materials, mathematical games have been integrated into those units that deal mainly with basic skills, in order to break the routine and increase motivation.

One of these units deals with beginning algebra, algebraic expressions, and substitution, and the accompanying game is called Steeplechase. One strategy for teaching this unit uses various modifications of the game appropriate to the students' progress through the unit from the opening lesson through exercising to mastery. The game can be modified to suit different ability levels.

Playing the Game

The game is designed for two to four players. It is played on a board with a track divided into strips containing an algebraic expression or an instruction (see fig. 25.1). Three piles of cards are placed in the middle of the board: eighteen blue cards on whose face the numerals 1 to 6 are printed (three cards of each numeral); eighteen yellow cards on whose face the numerals -1 to -6 appear (three cards of each numeral); and two green cards with 0 on the face. Each player has a marker.

The cards are placed facedown on the appropriate places on the board. Each player places his or her marker at GO. When their individual turn comes, players choose a card from the top of that pile they think may be most to their advantage. The player substitutes the number into the expression

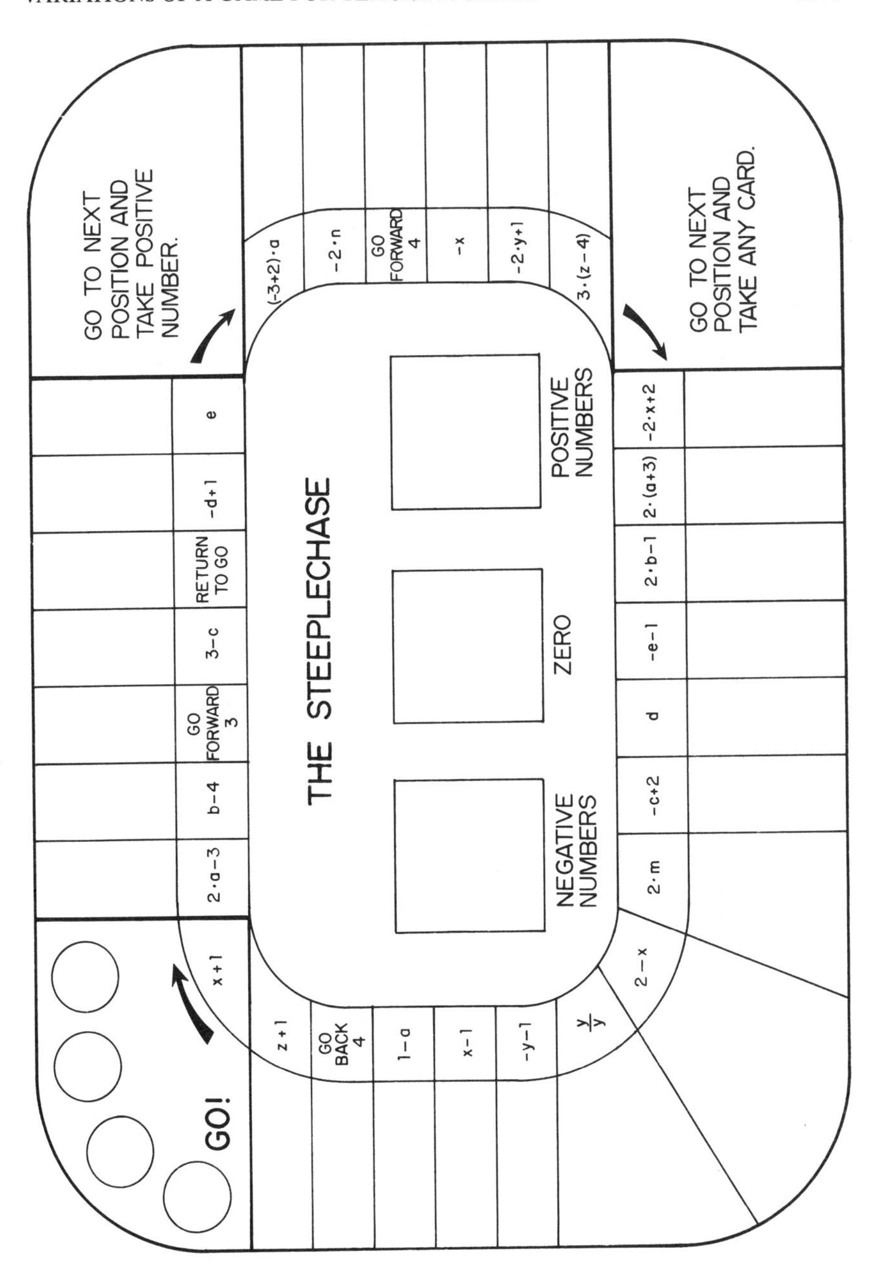

Fig. 25.1

written on the strip at which he or she stands, and the number obtained determines the move:

- *Forward* by that number of places if the number is positive
- *Backward* by that number of places if the number is negative
- *No movement* if the number is zero

When a meaningless expression is obtained, the player returns to GO. When players land on a place carrying an instruction, they obey it at once. The first player to go around the track twice wins the game.

An opening lesson

The main reason for developing the original game was to exercise basic algebraic skills (Friedlander 1977). The board for the original game carried "harder" algebraic expressions involving absolute values and fractions. Feedback from schools suggested that the game could be used for more than the exercise of basic skills and indicated that the game could be useful as an introduction to algebra. It could even be used prior to any formal teaching of the topic. In an experiment with the game used as an opening lesson, all the seventh graders in a school were given Steeplechase prior to studying algebra. They were told to substitute the number on the card they picked in place of the letter in the expression on the board. (The words *variable* and *open phrase* were not used.)

The students started to play with excitement, but soon a problem arose: "What is $2a - 3$?" "What is $-2n$?" The problem was solved by explaining that $2a - 3$ is $2 \cdot a - 3$ and $-2n$ is $-2 \cdot n$; the game continued. This minor problem in what was otherwise a successful beginning suggested the first variation on the original game: the multiplication dot to signify multiplication between the number and letter was added in all the relevant expressions on the board.

The insertion of the multiplication sign is often sufficient, but some students were found to have another difficulty altogether. For them some of the algebraic expressions were too complicated: they had difficulty with absolute value and expressions involving fractions. Thus the modified board shown in figure 25.1 was used.

In place of exercises

If the teacher begins with one of the variations described in the previous section (or with a conventional lesson-and-exercise scheme) and wishes to continue to exercise toward mastery using a game, then the original version is suitable for most students. However, brighter students who have already used the game as an introduction need a new challenge; otherwise, the game lacks sufficient motivation and thus defeats its own purpose. To change the algebraic expressions on the board to "harder" ones is neither sufficiently challenging nor relative to a reasonable educational objective. However, the rules of the game can be changed in such a way as to encourage the brighter students to plan their own strategy. For example, each student gets four cards, two with positive numbers and two with negative numbers. The remaining cards go facedown in their appropriate places on the game board. At each turn players must decide which card in their hand would serve them best. After substituting the number on their chosen card into the expression on the board and moving accordingly, players put this card on the bottom of the corresponding pile on the board and immediately draw another card from the top of the same pile to replace it. If they decide that zero would be their best choice, they may use more than one card from their hand if the sum of the numbers is zero. These cards are put at the bottom of the appropriate piles and new cards are taken from the top of the same piles to maintain their four-card hands. After a player has drawn from the top of a pile, play passes to the next player.

This version enables players to plan several steps in advance, which appeals to bright students and exercises strategic thinking.

Widening the field

Another possibility, which can be adapted for both strong and weak students, is to change the expressions on the board to two-variable expressions. This extends and enriches the topic at the same time as providing further exercise.

In this version, the game includes four piles of numbers—one positive pile and one negative pile for each variable. Each player chooses one card for x and one for y when his or her turn comes. The remaining instructions are unchanged.

The complication of having to consider the possible outcome of the substitution of two variables is sufficient challenge to make the game interesting even to bright students. For weaker students, the expressions need to be simple. The chance for further exercise is both desirable and welcome to such students, and the two-variable situation with simple expressions has sufficient novelty to maintain motivation.

Implementation in the Classroom

Three versions of the game were used in the experiment: as an opener, the original game with a change of rules, and a two-variable version. One lesson period was devoted to each version. In each group of players there was at least one good student. During the game, the teacher's role was confined to observation and guidance.

Student and teacher attitudes

The experimental group consisted of seven classes (representing different ability levels and socioeconomic background). In all seven a positive atmosphere developed, as well as an air of considerable expectancy from one lesson to the next and a great deal of student involvement. This impression of enjoyment and learning was reinforced in interviews with both teachers and students. The following student comments were typical:

- "The game helped me a lot in understanding the material, and I really enjoyed playing it. We all tried to solve the difficult exercises and to solve them alone."
- "It was fantastic! I didn't need the teacher to help me. I understood the game on my own and also the math through the game. I am sure that I can learn other topics without the teacher if there are more games, and I feel I understand better this way."

It was clear that the topic was regarded as one of the interesting mathematical topics by those who played the game, whereas the students who did not found it relatively uninteresting. However, the reactions were not uniform. Generally, high-ability students (but not from the socially deprived population) tired of the game quickly: "The game was interesting the first time, but after that I was bored. There was no challenge." Weaker students and students from the socially deprived population (irrespective of ability) enjoyed it more and more as they continued to play it. In addition, social relations developed in the classes that played the game. In particular, mutual student help appeared to increase.

The game's influence on achievement

The results obtained with regard to low-level skills, such as the mastery of substitution in algebraic expressions, were as follows. High-ability students (including the socially deprived) achieved mastery whether they played the game or not. Among the low-ability students, however, those who played scored considerably better on skills than those who did not. In particular, the socially deprived students not only achieved mastery but attained mean scores effectively identical with those of high-ability students of the non–socially deprived part of the population. As far as the higher cognitive levels are concerned, such as understanding and analysis, the following points stood out:

1. On the whole, the students who played the game did not regard the expression $-x$ as giving negative outcomes only, whereas nonplaying students did.

2. When asked to choose a number to substitute in a given expression, the students who played chose positive and negative numbers equally, whereas students who did not play chose positive numbers only. (Although the self-restriction to positive numbers was in greater evidence among the socially deprived, it disappeared to the same extent among those in the Steeplechase group irrespective of their social background.)

3. Given the number resulting from a (unknown) substitution, playing students succeeded to a far greater extent than nonplaying students in finding the number to be substituted to obtain the given result (that is, in performing the inverse process before they had learned to solve equations).

Conclusions

These findings would suggest that although Steeplechase was developed primarily to bring students to mastery in a basic skill, the intrinsic encouragement to think out the consequences of moves before they are made provides a considerable bonus. Because Steeplechase is integrated into the curriculum and demands specific mathematical activity at each move (rather than the general strategical thinking of many so-called mathematical games), specific mathematical cognitive achievement relevant to the students' immediate further studies results.

REFERENCE

Friedlander, Alex. "The Steeplechase." *Mathematics Teaching* 80 (September 1977): 37–39.

26

Organize a Math Games Tournament!

John C. del Regato

So you're thinking of organizing a mathematics contest—possibly one that makes use of educational games. That's most commendable. But there are some things you should know about planning and organizing such a contest. The following ideas may help, but they should not be considered rigid—only a framework for your own organizational efforts.

The major goal of a mathematics games tournament should be to bring parents, teachers, administrators, and students together in a social event that promotes the development of mathematical concepts and skills. To do this successfully requires careful planning around four major considerations: content, selection, leadership, and scheduling.

Content

One of the first considerations is *content.* Should the mathematics contest be based on puzzles, games, individual or team projects, or field events—or some combination of these activities? Puzzles, one-person games, offer students an opportunity to apply their knowledge of mathematics in a novel way. Written test competitions are a type of puzzle contest in which contestants attempt to solve as many puzzle problems as possible within an allotted time. These competitions can also be designed as team puzzles that allow participants to solve test items collectively. Team puzzles are noncompetitive games whose object is to establish cooperation among team members so that an identified goal can be achieved. Team puzzles differ from games in that they stress cooperation among participants to achieve an established goal, whereas games promote competition among players.

Mathematics contests can also center on individual or team projects to be judged by a panel according to some established criteria. Or field events can be the focus—staged to promote the application of contestants' mathematical knowledge or their ability to estimate. For example, a team can be asked to estimate the area of a given nonrectangular plot of land in a designated amount of time. Although calculators could be used, no actual measurements should be allowed. All these activities—puzzles, projects, and field events—offer many exciting possibilities.

Games are another possibility for a mathematics contest. As a convenient framework for discussing the important organizational considerations in mounting a mathematics contest, let us turn our attention to the specifics of organizing a mathematics games tournament.

Selection

What should be the criteria for selecting or developing games for the tournament? The school or school district, in cooperation with a college or university, may wish to select or design a special series of five to ten mathematics games for the tournament in order to promote the educational use of games. The games should be suitable for either home or school and should not require sophisticated equipment, elaborate procedures, or expensive materials. Games for tournament play should have well-defined rules. The winning goal needs to be clearly stated. If a tie is possible, it should be clearly stated in the rules. The mechanism for selecting a starting player should be established. The game rules need to be presented in no more than twenty-five numbered steps.

The mathematical content of games should be carefully considered. In middle school competition, for instance, the selected games should provide exciting challenges in computation, measurement, geometry, probability, and problem solving. Games involving prealgebra, algebra, geometry, statistics, or probability concepts will also eventually need to be evaluated on criteria involving the dynamics of play: suspense; skill and luck in equal

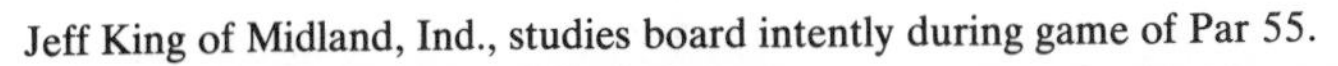

Jeff King of Midland, Ind., studies board intently during game of Par 55.

proportions; "repeat" play value, that is, the desire by players to "try it again"; a continuing competitive challenge, so that players can improve their mastery of the game; good action involved in play; nonverbal attributes; and family appeal.

Although the games should be geared to a designated grade level, they should have broad appeal. They should support and supplement the objectives of the general school mathematics curriculum at the indicated grade level. Some statement of the underlying mathematical concepts and skills being presented should be constructed. If the objective of a mathematics games tournament is to promote interest and enthusiasm in mathematics among students with a wide range of interests and abilities, the selected games need to be sensitive to the developmental differences of learners. In middle school, for example, not all students have achieved the stage of formal thought. Hence, games involving ratio and proportion or deductive logic should be carefully examined and field tested before being adopted. Some games that are desirable on a content basis may have a dampening effect on the interest and enthusiasm of the intended audience that can seriously jeopardize the major objective.

The games used in a tournament should provide immediate enjoyment, encourage spirited but friendly competition, offer opportunities for problem solving through the effective use of strategy, and emphasize participatory sportsmanship instead of winning. All the games should include an element of chance to broaden the possibilities of success. They should be as appealing to adults as they are to the participants. This is most important, since it will stimulate parents to help their children learn and practice the games for tournament play. Basically, this can be achieved by blending strategy with chance. Another result from such a blending is that no matter what a student's ability level may be, there is always a chance of winning. The games then become attractive to students at various levels of conceptual mastery and skill development.

Leadership

One of the greatest challenges in creating a mathematics games tournament is to locate leaders. It is important to include administrators, teachers, parents, and students and to group them in such committees as the following:

A *selection and design committee* is needed to test and evaluate games. They should carefully field test possible games before selecting the ones to be used. They need to draft a tournament manual, which should include information of the mathematical aspects of the games, the necessary materials, and a clear set of rules for each game. Tournament games for middle school students should have an average playing time of one hour or less. Game boards should be designed on 28-by-40.75-cm paper for easy reproduction at a printing center. Game boards of this size can easily be laminated, folded, and stored in an index file. Game directions should be available as an inserted sheet for easy reference. A reduced copy of the game board should appear in the manual of official tournament rules so teachers can make an overhead transparency for classroom discussion. Although the games may allow for more than two players, a two-person format is recommended for tournament play.

The *steering-organizational committee* is responsible for organizing and scheduling orientation sessions for parents, teachers, administrators, and student groups. These sessions should be scheduled at convenient times during the academic year prior to the tournament registration period. During these sessions, interested individuals should have the opportunity to learn the game rules through active play. The committee also recruits and prepares volunteers to officiate at the games on the day of the tournament. These officials can be administrators, teachers, teacher aides, other school district personnel, parents, and older students. This committee should coordinate the activities of the other committees and be responsible for implementing a successful tournament.

On the day of the tournament, the members of the steering-organizational committee should see to it that game tables are corded off so that only officials and contestants have access to the playing area. The tournament arena should be partitioned by the number of game events. Each event area should have sufficient space for several matches of the same two-person game to be played simultaneously. A large sign in the area of each event is helpful. Officials need to be on hand to help match players with the events indicated on their schedules.

A *communications committee* publicizes the tournament and announces the times and places for orientation sessions. American Education Week and parent-teacher functions are often good opportunities to promote a mathematics tournament. Student newsletters and state or local mathematics education publications can also alert interested individuals.

A *registration and scheduling committee* designs the registration form and distributes them to teachers or parents. Besides space for appropriate information on registrants, these forms should specify the time period for registration and the date and place of the tournament. They should also announce the orientation sessions, state any fees, solicit volunteers, and give the registrar's return address for submitting the completed form. This committee is also responsible for checking in contestants on tournament day and establishing a schedule for each player. A quota that provides for an equal number of male and female participants should be established. This internal mechanism offsets any sex bias that might be taking place elsewhere. Classroom teachers should be allotted, at most, one boy and one girl as representatives.

When registration is completed, a master schedule is prepared. All registrants should play each game in the tournament, and registrants should be carefully matched so that no two participants play against each other more than once. At check-in time on the day of the tournament, each player is given his or her own schedule of events including the time each begins and the name of the opposing player.

The *awards committee* attends to the purchase of prizes. All tournament players should be recognized for their efforts and should receive an award for participating. In addition to the awards given to all tournament participants, there should be performance awards—perhaps a first, second, and third prize. The committee designs and implements the awards ceremony, which takes place at the close of the competition.

Scheduling

The selection and design committee should have at least a year to collect or design, test, and select the games. Besides searching among commercial and teacher-made games, they should contact organizers of existing mathematics tournaments. These individuals are generally most happy to share their ideas on mathematics games.

Once the games have been selected and a manual of directions has been designed and written, the steering-organizational committee should be oriented to the tournament games and schedule from three to five orientation sessions for all interested educators and parents. The first of these sessions should occur in September or October, and the last is best scheduled in March or April. The tournament itself is best scheduled in May or June. Such a schedule allows ample time for teachers to become familiar with the games and orient their students. An orientation session specially designed for parents should be scheduled around the first of December.

All other committees should be formed within four months after the steering-organizational committee is formed. All members of these committees should receive a statement of their assigned responsibilities. The communications committee should announce all scheduled meetings at least two weeks prior to their occurrence.

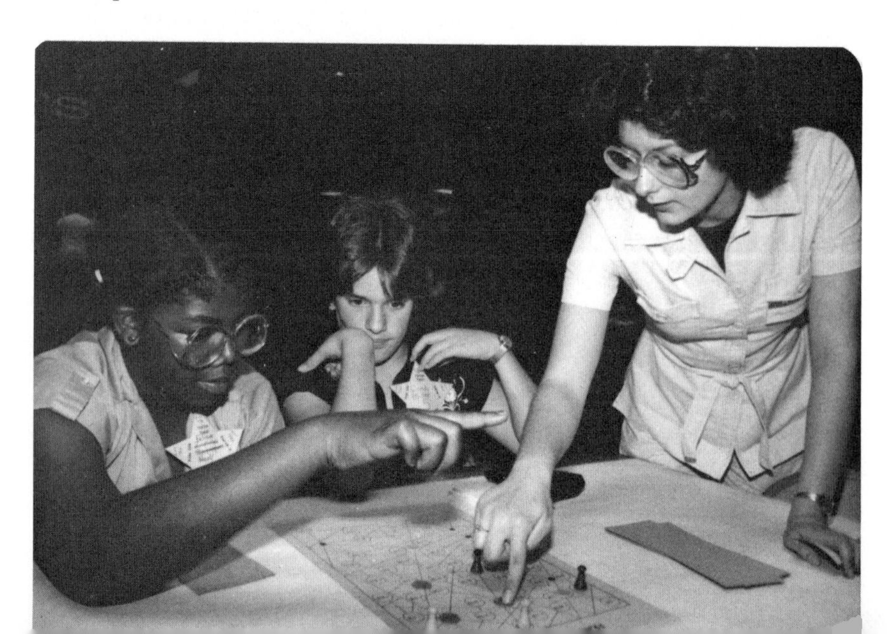

Within a few days after the tournament, an official list of all participants should be compiled and copies sent to each of the participating schools. It is advisable to send a letter of appreciation to all volunteers and cooperating teachers. A Tournament Hall of Fame can be initiated to recognize all the powerhouses who succeeded in achieving a first-place win in the tournament.

Many individuals will be affected positively by a mathematics games tournament. A variety of games that appropriately challenge students while giving them ample opportunity to experience success can make the tournament an exciting educational experience. Careful planning and coordination can encourage many members of the community to contribute their special talents and skills to this worthwhile endeavor. Parents who are looking for stimulating ideas for their children will want to become involved. Teachers and administrators who are looking for alternatives to paper-and-pencil exercises are potential tournament coordinators. Even students in high school or college may wish to volunteer their time as game officials. Thus, the players will not be the only ones to benefit. A mathematics games tournament is a most rewarding experience for all concerned and a powerful way to provide an interactive social experience that will enrich mathematics learning.

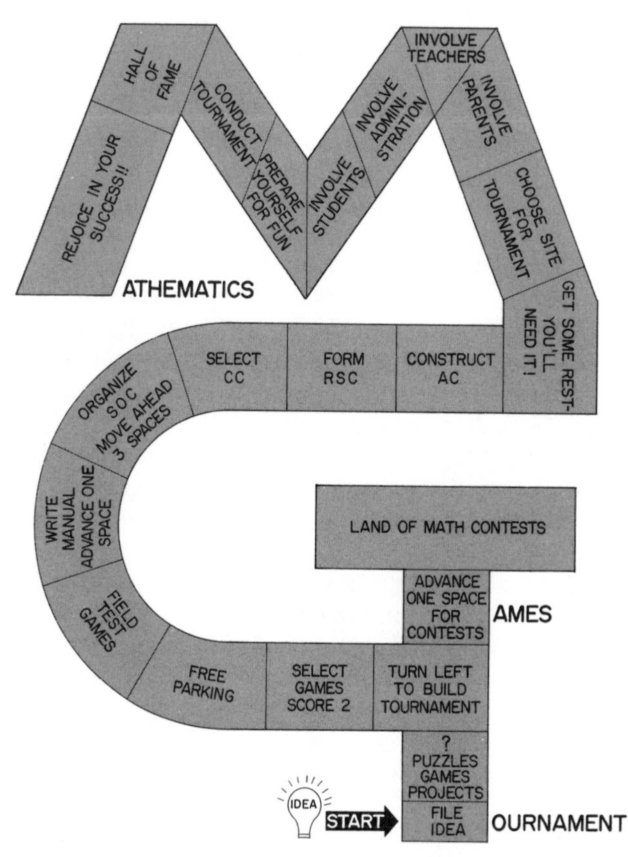

27

A Mathematics Carnival

Joan Michael Eschner
Betty J. Krist

A MATHEMATICS carnival is an activity well worth the class's—and the teacher's—time and energy. The original concept of a mathematics carnival was developed and executed by a graudate class of elementary and secondary school teachers at the State University of New York at Buffalo. The idea was rekindled with a heterogeneous fifth-grade class at Ebenezer Elementary School in West Seneca, New York.

Preparation

The preparation for the carnival began early in the school year. The teacher planted the seed by mentioning to the children the possibility of staging an activity that would allow them to show their parents and other students what they had learned about mathematics. At that time neither they nor their teacher had a clear conception of the final outcome.

During the months that ensued, the students kept a list of areas of study in which they had special interest. The teacher also recommended topics that appeared to hold promise for appropriate carnival activities.

A series of target dates was set up during the early planning stage (see Appendix A). It was also decided that the thirty students in the class could present ten different activities, each involving three children. An effort was made to cover ten different areas of mathematics and to allow more in-depth study of those areas.

The children then formed the groups. The most popular grouping principle was the idea of working with friends who shared common interests. Next in desirability was the natural cohesion of those students who shared common ideas. The formation of groups by the teacher on the basis of the strengths and weaknesses of individual children was employed only for those few students who had not already become group members.

By the end of the next week each group was expected to have an idea for developing their chosen area of study into a carnival activity. During this week the teacher witnessed many splendid things: a commitment to task, further research into chosen topics, the development of organizational skills, increased cooperation among students, and the fusion of curricular skills from art, English, science, and mathematics. The individual talents and abilities of each student brewed in a cauldron of creativity. All this did not, however, happen automatically; throughout this period the teacher scrutinized the students' activities, offering occasional suggestions or forecasting potential insurmountable obstacles.

Finally, the carnival's date was set. As the carnival evolved into its final shape, the students cut and painted boards, constructed games, and made cards, spinners, and awards. Some of the activities used electrical circuitry, which had to be completed and tested. It was interesting to watch how one idea bred another in the spirit of friendly competition. The booths grew from simple games to elaborate projects. The pageantry of a genuine carnival was developing. The children's enthusiasm grew as the carnival date approached. One week before the big day, the children wrote letters to their parents and to the other fifth- and sixth-grade classes of their school, inviting them to come and participate in the mathematics carnival.

The carnival was held in a regularly used classroom. Since the room needed to be partitioned, it was impossible to set up everything beforehand. Therefore, late in the afternoon of the day before the event, the students moved the desks to form relatively isolated booths. They then suspended from the ceiling large, painted cardboard dividers that served as signs or integral parts of the groups' activities. All was now ready.

Activities at the Carnival

The following morning the children, well rehearsed and excited, met their guests. One student at a time from each group was allowed to escort visitors to other booths, so that all in turn became involved in their classmates' activities. The visitors surveyed the wide assortment of activities and quickly began their tour of the carnival. The student hosts introduced the visitors to their classmates, who explained the activities of their booths. The visitors played the games, asked questions, and collected awards. Soon the noisy room was filled with enthusiastic children and adults who proudly displayed prizes awarded for winning a game. These were handmade paper badges with caricatures and colored ribbons.

Although other students in other schools should surely develop their own ideas in their own way, a description of a few booths can help to exemplify the flavor of the carnival and the kinds of activities that were enjoyable for these children (see Appendix B).

One group built a "Bowling for Decimals" booth, which consisted of soft-drink cans wrapped with paper and printed with decimal fractions. The cans were set up in a pyramid of six cans in front of a net at the end of a large table (fig. 27.1). The object of the game was to knock down as many cans as possible with one roll or toss of the ball. The numbers on the knocked-down cans were then added. If the player added the numbers correctly, he or she received an award. If the answer was incorrect, the students in the booth explained the error.

Fig. 27.1

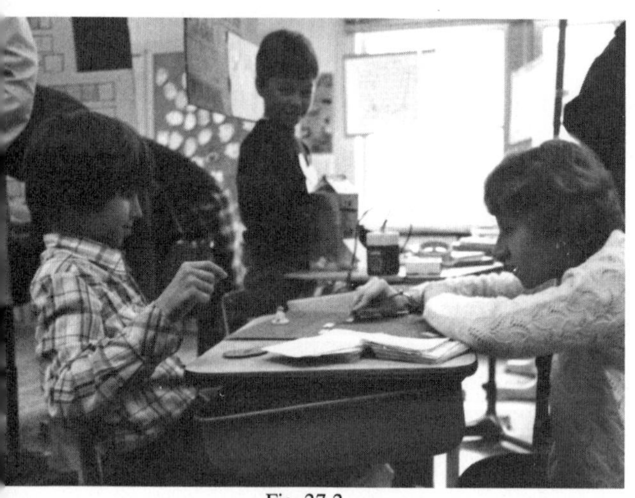

Fig. 27.2

Another interesting activity tested the participants' understanding of the metric system. "Metric Equivalents" featured a wired disk that had been marked off and labeled at an even number of regular intervals. The face of the disk was equipped with hands like those of a clock. The player was to move the hands so that they made contact with equivalent labels. A bulb would light when the player was successful. Figure 27.2 illustrates this game, modified because of electrical problems.

To play "Spin and Round" the player would draw a card from a deck. Each card was printed with an eight- or nine-digit integer. The player had to read the number and then spin a large spinner. Around the border of the spinner were terms indicating place values from ten to ten million. The player then had to round her or his number to the place indicated on the spinner.

"Penny Pitch" incorporated mathematical skill with physical dexterity. Cards were prepared with multiplication exercises that most children could do mentally. The answers to these exercises were glued next to paper cups that were countersunk in a large piece of cardboard. The player drew a card and mentally computed the answer to the exercise. The player then stood behind a line four feet from the board and pitched a penny, aiming for the cup that bore the correct answer (see fig. 27.3).

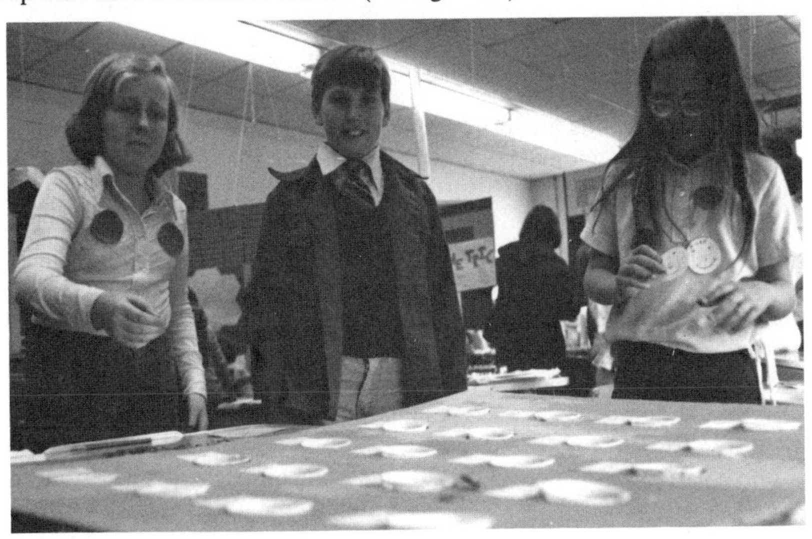

Fig. 27.3

One particularly ambitious student who had worked with negative numbers insisted on incorporating this concept in his booth (fig. 27.4). He soon learned many teaching skills himself, since most of the visitors to his group's booth, "Computer Patterns," had not been instructed in the nature of negative numbers. After teaching a sizable number of participants, he came to his teacher and said, "Now I know what it's like to be a teacher."

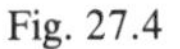

Fig. 27.4

Evaluation

The carnival was evaluated in an informal manner by the students. The overwhelming response from the participants was that it was fun. To a question about which aspects were considered to be "fun," one response was, "We had an idea that I didn't think was very good to begin with, but the more we worked on it the better it got. We never thought everyone would like it as much as they did. Even my dad said it was great." The children liked working with different materials, enjoyed the opportunity to build projects using skills they had learned from art, and derived genuine excitement and satisfaction from solving problems of design. Thinking was fun.

The children enjoyed being able to share the skills they had learned with the other fifth and sixth graders when they were explaining how their booths worked. They also enjoyed reteaching their parents, who had forgotten some of their mathematical skills.

The parents who attended the carnival were given a questionnaire that was to be returned the next day. The responses to the questionnaire reflected great support for this nonstandard way of teaching. Most parents indicated that they were glad to have an opportunity to be a part of such a positive school activity.

The teacher also evaluated this project from a pedagogical standpoint. The carnival approach is an excellent strategy for reinforcing concepts. Students must be sufficiently proficient in particular skills to see how they can be used in a game format. For students who only partially understand an idea, preparing for the carnival allows them to strengthen their understanding in a nonthreatening way. In addition, some children can successfully instruct their peers. Furthermore, the majority of the children considered several ideas involving a host of mathematical skills before making a final decision for their booth.

In many instances students had to learn additional skills before they could construct the booths. For example, one group of students found it necessary to divide a circle into twelve equal sectors. Learning to use a protractor had relevance for these students, and their booth was referred to when the protractor was a topic of instruction later in the school year. In fact, the carnival activities were frequently discussed in subsequent lessons.

To give the impression that such a project falls easily into place would be deceptive. As with any project, much planning, evaluation, and supervision is required. Many pedagogical issues must be considered. One potential pitfall in such an undertaking concerns the slower or less creative child. When a child does not have an original idea, it is difficult not to suggest a fully developed plan. But to do so would do an injustice to the child, for it would place him or her in a position of being unable to participate in the project comfortably and effectively. The teacher must be patient and perhaps reteach a concept to the students so that the mechanics of the booth are their own and they feel confident in their skills.

Another area that deserves special consideration is the time needed to allow the project to develop. Children need time to formulate their ideas, to experiment with them, and even to discard them yet still have time to develop an alternative plan. In spite of this, once the target dates have been selected, the schedule should be adhered to precisely.

An elementary and a secondary school teacher worked on this project together. They discussed the idea of this carnival before, during, and after its occurrence. An activity like this is a good opportunity for teachers of different levels to work together. Both students and teachers benefit from this kind of interchange of ideas.

Appendix A

Carnival Schedule

Early in school year: Teacher announces the possibility of having a mathematics carnival and gives students a general idea of what is involved.

Throughout the year: Students keep a list of the mathematics topics covered.

Mid-March: Students think of activities appropriate to a carnival.

End of March: Students form groups and choose topics. Groups report to teacher and get teacher's approval of topics.

First week in April: Groups develop suitable carnival activities related to their chosen topics.

Second week in April: Children report progress to teacher and set date for presentation of prototypes of booths.

Third week in April: Spring recess

Fourth week in April: Children refine ideas and present prototypes.

First and second week in May: Children work on equipment, prizes, and materials for booths. Groups work at different times and for different amounts of time as required by their projects. Students write and send invitations.

Second week in May: Carnival!

Appendix B

Carnival Booths

Bowling for Decimals

Metric Equivalents

Spin and Round

Penny Pitch

Computer Patterns: Player is given clues that describe the relation between lattice points and must identify those points.

Produce a Polygon: Player picks a card bearing the number of sides of a polygon and another card indicating its perimeter. Player must construct this polygon on a geoboard within the time limit of an egg timer.

Spin and Roll a Decimal: Player rolls a die and then spins a spinner indicating place value. Player must write the decimal value for the results of the spin and roll.

Reducing Fractions: Player draws a card containing a fraction that can be reduced. Player must then draw a picture representing the reduced fraction.

Multiplication Madness: Cards containing whole numbers and fractions are arranged in an array of pockets. A player chooses two cards, one from each of two pockets, and multiplies the numbers on them. Cards are color coded to show the level of difficulty.

Dazzling Divisions: Many posters illustrating the division of fractions are displayed. Player picks a poster and must name and solve the problem represented by that poster.

28

A Mathematics Fair

Judy Day
Joan McNichols
Jeanne Robb

WHEN a mathematics fair is approached with goals and objectives in mind, it can be the key to making mathematics meaningful to middle school children. Maria Montessori (1967, p. 152) said, "We put it like this: The child's intelligence can develop to a certain level without the help of his hand. But, if it develops with his hand, then the level it reaches is higher." A well-planned mathematics fair capitalizes on this philosophy.

The middle school mathematics fair encompasses exploring, investigating, hypothesizing, experimenting, and formulating generalizations. The students are involved in doing mathematics, manipulating objects, discussing, writing, thinking, and solving problems as well as learning skills with which to communicate to their peers and families the information they gather. The teachers and mathematics specialist become catalysts and resource people, invariably getting involved in the investigation side by side with the students.

Another major plus offered by the mathematics fair is the noticeable improvement in students' problem-solving and problem-posing skills. Abstraction at this level is almost never mastered in the basic arithmetic program. The traditional mathematics program is effective in getting learners involved in problem-solving techniques but often does not involve them sufficiently to bring them to the problem-posing stage, a more abstract level of thinking. The mathematics fair can involve all three skills of computation, problem solving, and, finally, problem posing.

Materials

The activities for this particular project were adapted from a series of activity cards presenting a wide range of topics and skill levels (Hewitt 1967). Professional journals and resource books could also be used as sources of activities. Sample activity cards are shown in figures 28.1–28.4.

HOW FAR AROUND

- o - o - o -

You will need string, scissors, a pen, and several cylindrical objects of different sizes. (Containers with straight sides are good.)

1. Put the string around your container, making sure it goes straight around the container, not slanting at all. Let the ends of the string overlap. Make a pen mark on the two parts of the string where they overlap.

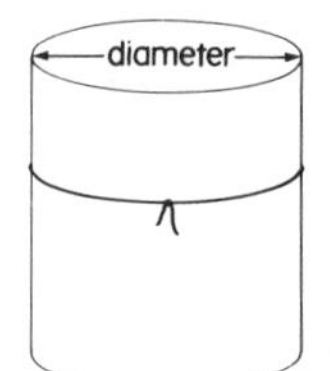

2. Remove the string and cut across the marks. You now have a piece of string as long as the distance around the container; this is its **circumference.**
3. With another piece of string measure the **diameter** and cut a piece of string exactly this length.
4. Prepare a chart like this.

	Circumference (C)	**Diameter (D)**	**Number of times C is greater than D**
First container	cm	cm	

5. Put the string that measures circumference on your desk and see how many times the diameter fits across it. There should be a bit left over. Approximately what fraction of the diameter is this?

Circumference

Diameter

Record this result in your chart.

6. Do the experiment at least five times with containers of different sizes. **What did you discover?**
7. Use the measurements from your chart to make a graph of your results.

Fig. 28.1

Organizational and instructional plan

The teacher's preparation involves preparing appropriate activity cards and grouping the children in pairs so that partners have similar ability. Each pair is given an activity that will be challenging but not overwhelming.

The children are introduced to the fair in a total group presentation that includes rationale, a general explanation, and a discussion of how the students are to go about presenting their projects. This presentation is given to two separate classes, with the explanation that each pair will have a counterpart in the other class working on the same project. Keeping the two pairs separate in the beginning forces all the children to become actively involved.

PYTHAGORAS
and the Right-Angled Triangle

A Greek mathematician named Pythagoras discovered some interesting facts about right-angled triangles. Historians say he noticed it as he was looking at a tiled pattern on a floor. This activity will help you make the same discovery.

1. On a large sheet of paper draw a right-angled triangle with sides 3 cm, 4 cm, and 5 cm.
2. Put a row of centimeter squares on the outside of the triangle. Now build each row outward until you have a square on each side of the triangle.
3. Count and record the number of centimeter squares in each large square.
4. Do you notice anything about the number of centimeter squares on the two short sides of the triangle and the number on the long one?

Challenge

1. Build another right-angled triangle on the **far side** of the large square with edges 5 cm long. Keep the other line segment that forms the right angle 3 cm long.
2. Follow steps 2, 3, and 4 above.
3. Build another right-angled triangle on the **far side** of the largest square. Always keep the other line segment that makes the right angle 3 cm long. Does the pattern continue when you build squares on the sides of the triangles?
4. Predict the shape you will get as you build more squares on the line segments of the triangles.

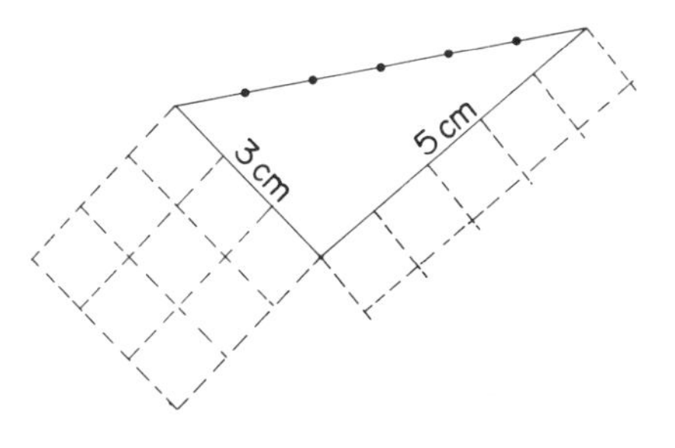

Fig. 28.2

Some overlapping of ideas will occur, but the children tend to generate more creative ideas when they begin working in smaller groups.

Four key points that should be discussed with students at the initial presentation are these:

 **ROUGH GOING** 

or

Noting the RESISTANCE of Surfaces to a Moving Object

1. Make a list of all surfaces that you think are smooth or fairly smooth for moving on.
2. Can you think of any surfaces that make moving difficult? Make a list of these.
3. Collect a number of materials with different surfaces, such as linoleum, sandpaper, carpet, or glass. You will also need a small toy (e.g., a car), a piece of smooth wood about 30 cm × 60 cm, a ruler, and a weight, such as a heavy book, to hold the wood in place.

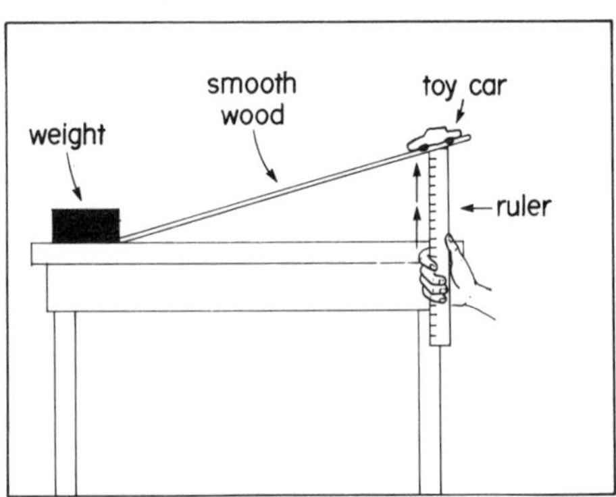

4. Place the car on the wood.
5. Hold the ruler vertically under the wood and slowly push the wood up.
6. Have your partner watch very carefully, and say STOP as soon as the car begins to move.
7. Measure how high you raised the wood.
8. Repeat the experiment putting different surfaces on the wood under the car.
9. Make a list of your results.
10. What conclusions can you make as a result of your experiments?

Fig. 28.3

1. How to go beyond the actual questions posed on their activity cards, look for depth, and pose questions of their own
2. How to enhance their presentations with such visual aids as posters, models, and so on
3. How to speak effectively before a group (including such points as how to budget their speaking time, use notes, and speak clearly)
4. How to choose an activity that will demonstrate the concept while actively involving their audience

Now the children can be given their group assignments and activity cards

FRICTION

A word that can be used instead of **resistance** (when we are talking about the resistance of surfaces to a moving object) is **friction.**

1. Make a list of the surfaces you used in the "Rough Going" experiments.
2. List them in order, beginning with the surface that made the least resistance to the toy car. How do you know which one made the least resistance?
3. Look at your results. Can you find suitable words to make this statement true, based on your experiments?

The __________________________ a surface is,
the __________________________ resistance it offers to a moving object.

Friction All Around

Sometimes friction helps and sometimes it hinders!

1. How is friction reduced on a dance floor?
2. How is friction increased on icy roads in the winter?
3. List situations in the world around you where friction, or the lack of it, is important.

Fig. 28.4

and begin their research. The pairs are given approximately two weeks to develop their ideas. During this time the teachers should be available as advisors for about five class periods.

At the beginning of the third week the groups are united with their counterparts, and they begin collating information and working out the mechanics of their presentations. Posters, models, and visual aids begin to appear. This process involves another two weeks. It is important to establish a calendar of presentation dates when the two groups first come together. Each group of four is assigned one forty-five-minute class period on a given date. All students receive a copy of the calendar, and they are encouraged to invite family and friends to their presentations.

This month of exploring, questioning, and creative thinking highly motivates the students, and they are enthusiastic about beginning their actual presentations.

Conclusions

Visual students are being educated in schools that are heavily geared toward the auditory learner. A mathematics fair gives these visual learners (who make up 30 percent of our student population, according to some sources) the consideration they need (Barbe and Milbre 1980). Children with mathematics anxiety are reached by being put to work on a meaningful project they can handle and by being given a less anxious partner with whom to work. The final rewards of accomplishment increase the children's self-confidence about their ability in mathematics, another necessary ingredient to teaching a child. Visual learners find this approach ideal because it allows the movement, experimentation, and discovery that is their style of learning. Auditory learners, although very comfortable in the traditional learning situation, find their needs being met in the presentations and discussions that follow the actual research and exploration. This multifaceted effect must be taken into consideration, since the research demonstrates that the traditional mathematics lesson reaches only a small percentage of students. This approach, as one part of a total program, accommodates many kinds of learners.

An evaluation of past fairs has disclosed some common threads.

1. Students, regardless of ability, need to go through concrete activities before being expected to grasp more difficult levels of abstraction. The process of *doing* mathematics seems to be as valuable as the knowledge gained. Out of this active participation, students are far more likely to see patterns and order in mathematics and are able to verbalize relationships between one idea and another (Maletsky 1972). Understanding comes through intense involvement with the problems as the students are guided into thinking for themselves while they participate in the problem-solving situations. Problem posing often follows the initial problem-solving task, as in the experience of a group who had to build a model to demonstrate the

Pythagorean theorem. Having successfully executed the model with multibase materials and refined their results to more accurate measurements by reproducing their squares of the sides of the triangle on graph paper, they then embarked on investigating the possibility of building further squares on further triangles. Ultimately they discovered that the sequence of square numbers and the irrational spiral can be generated in this manner.

2. The tremendous enthusiasm and vitality produced by mathematics fairs is contagious. One group's exciting discoveries of novel ways to interpret and communicate information inspired ideas for other groups. Everyone wanted to improve on the previous presentations. There was enough flexibility to allow follow-up activities to flow over into other areas of the curriculum in the form of crosswords (vocabulary), artwork, history, and library research skills.

3. This approach appears to foster self-esteem, especially in those children who generally do not experience success in a traditional arithmetic (computation) lesson. Building models seems to produce confidence because individuals are able to work at their own rate and can use the models to check their work frequently. These children are less threatened by failure and more likely to achieve success, since each step can be seen to be right or wrong as it is carefully completed.

4. Myriad opportunities exist for children to improve their communication skills and for teachers to observe children at work. Valuable insights can be gained from noting each child's attitude and pattern of learning.

5. It takes time for each group to become comfortable with the project.

Also, when teachers talk less in the initial stages, more unusual discoveries can result, since students often see dimensions not apparent to adults.

6. When certain topics appeal to the whole group, subsequent lessons can be planned around these concepts. Some of these lessons afforded an opportunity to correlate science and mathematics. A project on estimating led to work on the ESS (1975) unit "Peas and Particles" and to different ways of finding animal populations in the environment as outlined in the OBIS (1975) Trial Module.

7. The gains from a mathematics fair flow over into everyday mathematics lessons. First-hand experiences of measuring length, perimeter, area, and volume, seeing symmetrical objects, finding patterns, and so on, enhance the children's view of the breadth of mathematics. Even students who have not been previously exposed to a concrete approach are surprisingly creative.

The passive learner, according to Jerome Bruner, does not do as well as the active learner (Tobias 1978). Alan Natapof, a physicist and researcher of the structure of the brain, believes that mathematics is something people *do*, and that to do it well they must do it often and intensely. He proposes that mathematics activities must use all the senses if those activities are to be truly learned and if the brain is to make mathematics part of its repertoire (Tobias 1978). This idea implies using the body more than is usual in a typically sedentary mathematics class. The mathematics fair accomplishes this.

These researchers are not alone. The topic of "mechanical" (rote) versus "meaningful" (activity-centered) learning has long been debated. When computing skills are the desired result, then either method will probably suffice. However, when retention is the criterion, the meaningful method outranks the mechanical, for it appears to produce not only greater transfer but also increased understanding of the principles of mathematics (Travers 1973). This principle is exemplified in the project of the mathematics fair.

REFERENCES

Barbe, Walter B., and Michael Milone. "Modality." *Instructor* 89 (January 1980): 44–47.

Elementary Science Study Program (ESS). "Peas and Particles." New York: McGraw-Hill Book Co., Webster Division, 1975.

Hewitt, B. *Guide to Adventure in Mathematics.* Hayward, Calif.: Activity Resources, 1967.

Maletsky, Evan M. "Aids and Activities." In *The Slow Learner in Mathematics,* Thirty-fifth Yearbook of the National Council of Teachers of Mathematics, pp. 182–220. Washington, D.C.: The Council, 1972.

Montessori, Maria. *The Absorbent Mind.* Translated by Claude A. Claremont. New York: Holt, Rinehart & Winston, 1967.

Outdoor Biology Instructional Strategies (OBIS). Sets 1 and 3. Berkeley, Calif.: University of California, Lawrence Hall of Science, 1975.

Tobias, Sheila. *Overcoming Math Anxiety.* New York: W. W. Norton & Co., 1978.

Travers, Robert, ed. *Second Handbook of Research on Teaching.* New York: Rand McNally, 1973.